THE WORLD'S DIN

THE WORLD'S DIN

Listening to Records, Radio and Films in New Zealand, 1880–1940

PETER HOAR

OTAGO

Published by Otago University Press
Level 1, 398 Cumberland Street
Dunedin, New Zealand
university.press@otago.ac.nz
www.otago.ac.nz/press

First published 2018
Copyright © Peter Hoar
The moral rights of the author have been asserted.

ISBN 978-1-98-853119-9

A catalogue record for this book is available from the National Library of New Zealand. This book is copyright. Except for the purpose of fair review, no part may be stored or transmitted in any form or by any means, electronic or mechanical, including recording or storage in any information retrieval system, without permission in writing from the publishers. No reproduction may be made, whether by photocopying or by any other means, unless a licence has been obtained from the publisher.

Editor: Jane Parkin
Indexer: Diane Lowther
Design/layout: Fiona Moffat

Cover: Violinist with Cylinder Phonograph Collection, Manawatu Heritage, Ian Matheson City Archives, 2015G_Young347_010410

Printed in Christchurch by Caxton.

CONTENTS

	ACKNOWLEDGEMENTS	6
	PREFACE	
	Hearing the Past: A Sound History of New Zealand	7
	OVERTURE	
	Listening to Recorded Sound	11
	PART I RECORDS	
CHAPTER 1	*Exhibiting Sounds*	21
CHAPTER 2	*Selling Sounds*	34
CHAPTER 3	*Domesticating Sounds*	44
CHAPTER 4	*Teaching Sounds*	53
CHAPTER 5	*Moving Sounds*	67
	PART II RADIOS	
CHAPTER 6	*Receiving Radio*	87
CHAPTER 7	*Military Radio*	104
CHAPTER 8	*Hearing Radio*	119
CHAPTER 9	*Organising Radio*	136
	PART III FILMS	
CHAPTER 10	*Living Pictures*	155
CHAPTER 11	*Orchestrated Pictures*	171
CHAPTER 12	*Talking Pictures*	184
	CONCLUSION	200
	NOTES	207
	BIBLIOGRAPHY	244
	INDEX	278

Acknowledgements

A book is a communal project and I am grateful to all the people who have helped me along the way.

This book began life as PhD thesis, so first thanks go to my supervisors Associate Professor Caroline Daley and Dr Joseph Zizek along with the rest of the staff at the University of Auckland's History Department.

Many thanks also to all my colleagues at Auckland University of Technology for their support, encouragement and comradeship.

Research would be impossible without dedicated and committed librarians and archivists. Much gratitude and plaudits to all the individuals and institutions that collect, organise, index, catalogue, preserve and generally maintain the raw materials of our histories.

I'm very grateful to Otago University Press for taking this project on board. Thanks to Rachel Scott, Imogen Coxhead and Fiona Moffat for their patience, guidance and efforts. A very big thanks to editor Jane Parkin for her peerless work on the text. I had a very good team to work with but of course, any mistakes are all mine.

A really, really big 'thank you' – in upper case, 30-point font and with fireworks too – to my friends and family for all their support, kindness, laughs and love. Finally, my heartfelt thanks to Robyn, for being there with me.

Preface

HEARING THE PAST: A SOUND HISTORY OF NEW ZEALAND

• • •

The Marbecks music shop in Auckland's Queens Arcade was established in 1934, and became well known for its varied stock and knowledgeable staff. Roger Marbeck, grandson of the store's founder, recalled that the original shop had a piano and 'sold sheet music. Then there was a gramophone and then records and later it was cassettes and CDs'.[1] When it was announced in November 2012 that the store would close the following March, there was sadness but not much surprise. Record stores worldwide have been closing in the face of competition from online services such as Amazon, Play and iTunes, along with practices such as file sharing across the internet and CD copying.[2] 'It's just the way things go,' Roger said. Changes in audio technologies had, it seemed, spelled the end for one of Auckland's most loved musical institutions.

Happily, it was not the case. Marbecks reopened in Queens Arcade in 2014 and is still going strong.[3] It sells CDs from the store and online. It sells a lot of vinyl records too, and has a window display devoted solely to the format, as well as several large bins full of LPs in their colourful sleeves. Vinyl records were meant to die off when the compact disc was developed in the early 1980s, but have made a significant comeback over the last few years.[4] Nostalgia, audio quality, snobbery, tangibility, fashion, cultural capital, community and the joy of the hunt for LPs are just some of the reasons given for their resurrection.[5] Vinyl fans often argue that the medium delivers a better, 'warmer' sound than the CD or other digital formats. Canny marketing and events such as Record Store Day have also played major roles in the revival.[6]

At the heart of these assertions are differing concepts about the nature of listening and how it is shaped by social, cultural and personal factors. Frequency ranges and other qualities of audio technologies can be measured and charted, but it is quite another thing to objectively quantify concepts like 'warmth'. Listening to audio machines is not a science, and the vinyl revival shows just how culturally loaded and nuanced our ways of hearing the world can be.

Since 1877 recorded sound has been stored on tinfoil, wax cylinders, shellac discs, magnetic tape, vinyl records and CDs. These are all tangible artefacts. They take up space, and can be handed around, scratched, drawn on, smashed or carefully preserved as vessels of personal memory. They are also objects that fundamentally altered human experience by disembodying sounds from the people who made them. Digital technology takes this disembodiment even further by turning sounds into combinations of ones and zeros that form the binary code of software. Recorded sound is now weightless and massless, and can be instantaneously moved around the world via the internet. The world is there for the hearing by anyone who has a computer or a phone.

The digital moment might seem to be a revolution in the ways we listen. We have no more need for objects such as records or CDs; audio files can be stored in the cloud or on an iPod. Radio stations now have to compete with services like Spotify, iTunes or Pandora that allow listeners to hear continuous music of their choice with or without advertisements (in that typically modern twist, we now pay extra to be spared the overt assault of commercialism). Films can be experienced on large screens at home with high-quality sound systems, or on laptops or tablets wherever the viewer happens to be. But these digital technologies are extensions of the audio machines that have gone before – and in vinyl's case have come back again. The real shock of the sonic new happened between 1877 and the late 1930s.

The experiences of sound through the technologies discussed in this book were part of the disruptive experience of modernity. Film scholar James Lastra describes it as 'an experience of profound temporal and spatial displacements, of often accelerated and diversified shocks, of new modes of society and of experience' that has been shaped by media technologies.[7] Gramophones, radios and films appear to fragment perceptions of time and space through sound. Sounds are separated from their sources through recordings and transmission. This has added a new dimension to human experience. Before the development of machines such as the phonograph, sounds had been heard only in the vicinity of the sounder.

The iPod is related to the portable gramophone in terms of listening practices: both are used by listeners in creative ways to experience the spaces and sounds of modernity. The iPod is not a revolution in itself; it is a refinement of the technology that captured, stored and replayed sounds which was developed by Thomas Edison and others during the later decades of the nineteenth century. Modern radio and cinema practices, too, may seem completely different from their historical antecedents, but they are also refinements of earlier technological developments.

Records, radios and films freed sounds from the constraints of bodily presence and radically altered traditional ideas about sound and listening. They transformed notions about performance and accelerated the commodification of music. The sounds heard through wax cylinders, shellac discs, films and crackling radio receivers generated new cultural hierarchies and ideas about appropriate listening that still resonate through contemporary culture and life. These technologies also changed educational methods and practices. They turned the production of sound into a global industry and altered leisure habits. New forms of music such as ragtime, jazz and hillbilly were spread around the world by and through audio technologies. Traditional ideas about music, sound, noise, and who gets to determine just what these are, were undermined.[8] Allowing the dead to speak or sing, and the living to be heard at a distance, altered notions of time and space and sharpened questions about what it is to be human.[9] These strange, disorientating, sometimes disturbing and exciting effects were heard and felt in New Zealand as they were worldwide.

The sounds of history in New Zealand have received very little attention from historians and other scholars. Discussions of recording, radio and cinema tended to focus on the fortunes of institutions and individuals rather than how these technologies sounded. There has also been a nationalistic focus on making sounds in New Zealand, rather than on New Zealanders hearing the sounds of the world. The notion of 'hearing the world' is in tune with the challenge offered by New Zealand historian Peter Gibbons, who suggested that New Zealand's historians needed to move beyond narrow ideas of national identity and pay more attention 'to the world's place in New Zealand'.[10] This would call attention to the 'convergences of experiences in these parts of the world with experiences of peoples in other parts of the world'.[11] These experiences include listening.

The idea of 'hearing the world' connects the rich acoustic tapestry of everyday life with the global patterns of industrial production and consumption of canned sounds. Understanding how popular culture was heard in New Zealand is a way of recovering some of the richness and variety of past everyday life. Gibbons' idea is particularly relevant for this book given that the great majority of the recorded sounds heard here have been from other places, mainly the United States. Concentrating on films or records made in this country ignores most of the sonic cultures historically experienced and enjoyed by New Zealanders.

Over the last three decades, historians have opened their ears to the idea that acoustic environments have pasts and the past had acoustic environments. There are now important historical studies about sounds in early modern England, revolutionary Paris, the countryside in nineteenth-century France and early twentieth-century America, to name a few.[12] But there are no such studies for New Zealand. General histories such as those by Keith Sinclair, James Belich and Michael King, for example, make only passing references to audio technologies and do not explore what it meant to hear these machines.[13] Such histories have emphasised, and often celebrated, the uniqueness of New Zealand culture and development.[14] In these accounts, the idea of the New Zealand 'nation' and its unique culture are dominant tropes that elide the complex and ongoing relationships between the rest of the world and local culture.[15] Such a historiography also ignores the willing and active participation of New Zealanders in the global consumer and leisure cultures associated with modernity and technology.[16]

Beyond the idea of broadening scholarly horizons, ignoring the roles of sound in New Zealand's cultural history is like watching a film with the sound turned down. We need to turn it up and bring on the noise. This noise can sometimes be annoying; it can disrupt and confuse. But listening to the sounds of history can also be rewarding: they encourage us to think about the past in new ways; they reinforce some of the revisionist interpretations of New Zealand history, especially in terms of leisure, consumerism, modernity and globalisation; and they are, in and of themselves, enjoyable. Sound historians do not replay, they remix the records of history.

Overture

LISTENING TO RECORDED SOUND

*The world is not for the beholding. It is for hearing.
It is not legible, but audible.*
JACQUES ATTALI[1]

I have a modest but growing collection of 78rpm records. One of my favourite discs is 'Tip-Toe Through the Tulips' sung by Nick Lucas, who was also known as 'The Crooning Troubadour'. The record was made in 1929 and was a hit from the film *Gold Diggers of Broadway*. Hearing this record is by no means an easy process. I need to open up my portable gramophone, which is a Columbia Graphonola made some time during the 1930s. I have to unscrew the old needle and replace it with a new one. Then I have to turn the handle until it locks into place: too much winding will break the spring and too little will make the record play slowly. After releasing the turntable lever, I lower the needle onto the edge of the record, and it catches the groove inscribed on the disc in a spiral from the outer edge to the inner. The needle vibrates as it follows this trench in the disc. These vibrations are transmitted to a membrane inside the gramophone, which causes vibrations in the air that are then amplified by a horn built into the machine. The sound waves are focused by my outer ear onto my eardrum that in turn moves the bones in the middle ear called the malleus (hammer), incus (anvil) and stapes (stirrups). These rhythms then move tiny hair-like projections inside my fluid-filled cochlea (stereocilia), which in turn convert the movements into electrical currents that the auditory nerve conducts to my brain's auditory cortex. This is hearing.[2] When the biochemical processes in my brain then give rise to music in my consciousness, I am listening and the world becomes audible.

I could hear 'Tip-Toe Through the Tulips' from a vinyl record, a cassette, an eight-track cartridge, a magnetic tape reel or a CD. I might click an mp3, or stream it from Spotify or iTunes. I can hear (and see) it on YouTube.[3] Or I could watch a DVD of *Gold Diggers of Broadway* and relish the bass on my home theatre. A local cinema might be showing the film so I could hear the song filling a theatre. I might be able to hear it on a radio. All these technologies allow me to hear the song in different places and at different times and as often as I want – and each of them affects how I listen to it.

Before the development of the phonograph in 1877, all sounds were ephemeral. They were made and then lost. Music could be notated and spoken words written down, but the sounds themselves were gone forever as they faded into inaudibility. Recording sounds fundamentally altered our relationship with our world and our perception of it. Sounds could now be slowed down, sped up, amplified and subjected to scientific scrutiny. They could be manufactured and distributed on an industrial scale through studios and recording companies. They could be manipulated to make new audio experiences, just as photographic technology could be used to mask the world as much as to display it. This was a huge revolution, and it is too soon to describe its ultimate effects. But all our modern audio technologies are in many ways simply refined versions of the audio recording and transmitting technologies that originated between the 1870s and the 1920s. This sonic singularity shattered traditional ideas and experiences of listening.

Where once a sound could be heard only once, in one place, and only by those present, records, radios and films allowed sounds to be heard repeatedly, at any place and by anybody. The techonologies even seemed to conquer death in allowing the dead to continue to speak and sing.[4] Sound became portable, durable and repeatable. These three themes are important leitmotifs that resonate throughout this book.

Radio transmitted sound across vast distances. Recordings could be heard anywhere there was a suitable player. This portability was used as an important selling point by early twentieth-century recording companies like Decca, whose advertising slogan was 'She shall have music wherever she goes'.[5] Radios brought a torrent of varied music and voices into homes. Films offered new music to people in theatres, town halls, schoolrooms and wherever else a peripatetic projectionist might set up. In 1929 Nick Lucas might have been heard singing 'Tip-Toe Through the Tulips' at a cinema, on a gramophone at the beach and on a domestic radio all in the same day. Sound had become portable.

Gramophones and radios had become common features of the domestic spaces of New Zealand by the late 1930s.[6] They replaced the piano as the typical domestic music machine, and in so doing raised questions about ways of listening. In public places the etiquette and modes of listening were, and are, often enforced by audience members to ensure conformity to notions of correct behaviour and deportment when hearing 'serious' music.[7] The

LISTENING TO RECORDED SOUND

Never mind the game, let's have a cool drink and listen to some music. Being modern meant having music wherever you were.
Evening Post, 26 April 1924, 24

gramophone and radio, although they could be turned to commercial ends, challenged accepted ways of listening not only in public but also at home. Was it appropriate to insist on a reverential hush when hearing a Beethoven piano sonata on a gramophone positioned on a front porch? Did dance music issuing from a radio in a parlour call for the sorts of listening and bodily movements found in nightclubs? The answers to such questions were sometimes 'yes' and sometimes 'no'. People used and heard the machines as they saw fit. A gramophone might be employed to study music at home at one moment and the next be supplying music for a dance party. Depending on the context and situation, a radio broadcast might be heard attentively or treated as background noise.

The new visual art of cinema also offered new spaces in which to hear music. Specialised theatres for film-showing began appearing in New Zealand during the early 1900s.[8] While the sounds heard with the films were usually made by live performers, gramophones and phonographs played large roles in these new public spaces even after the advent of the talkies (films with synchronised sound) during the late 1920s. These spaces, too, raised the question of how to listen. Were audience members to remain attentive and silent as though at a recital of classical music? Were they allowed to supply their own sounds in the form of interjections, comments and singalongs? How were these new entertainments, seemingly alive and yet dead at the same time, to be heard? Film historians have pointed out that during the so-called 'silent film' era audience members expected to hear music and voices with the films. It is something of a cliché that silent films were never really silent. A wide range of sonic practices were typically part and parcel of cinema-going before the advent of synchronised sound.[9]

Portability made sound inescapable. Records and radios brought new sounds, sometimes unwelcome, into the private space of the home. Public spaces such as theatres, the streets, shops, the countryside also became saturated with sound. The world's sonic environment was radically transformed by mechanically stored or transmitted sounds.

The second major change associated with recordings, radios and films was that sound became durable. Sheet music and scores preserve instructions for making sounds rather than the sounds themselves. Mechanical instruments, such as music boxes and player pianos, store sounds but only the few they are programmed to emit. The phonograph opened up the possibility of preserving any kind of sound, anywhere, at any time, and this capability permanently altered listening. A recording was a possession that could last and be handed on like the family china.

In the nineteenth century all music was live, so the sonic qualities of a performer such as singer Jenny Lind exist today only in the form of written descriptions.[10] However, the voice of an artist such as the opera singer Enrico Caruso, who left a body of recordings, can still be heard many years after his death. The durability of recordings allows scholars to analyse the styles of earlier performers, and trace changes in musical techniques and mannerisms.[11] This was a major factor in the transformation of musical aesthetics during the early decades of the twentieth century.

Musical education, too, was transformed by the permanence of recordings. Learning through imitation became very important in the age of technology. Many jazz musicians in America, for example, learned songs and playing styles from repeated listening to records.[12] Films also provided useful visual guides to performance techniques. Through recordings, New Zealand musicians learned how to play jazz, blues, country and other music that written scores could not entirely capture. They may have been far away from the bars, brothels and speakeasies of New Orleans, Chicago or Nashville, but they were able to reproduce the music of those locations. The durability of recordings underpinned the global spread and dominance of American popular music and culture during the twentieth century.

Anthropologists and ethnomusicologists were quick to grasp the opportunities for preservation offered by recording technologies. They built

Record companies such as the short-lived Durium label used durability as a marketing point. These records were made of resin and did not break as easily as the standard 78s made from shellac. Evening Post, 9 June 1932, 5

collections of sounds.¹³ These sonic museums can be valuable for the recovery of long-deceased musical folkways, but they are not without their hazards. Recordings frame songs, ceremonies, chants and poems in certain ways. Wax cylinders and acetate discs held, at most, only six minutes of sound. Recorded performances were made for the machines themselves, lifting those performances out of the contexts in which they were normally experienced. For example, the recordings of Māori music made by Percy Grainger during his 1909 tour of New Zealand remain a valuable source of musicological information but do not reflect the ways in which contemporaries experienced that music 'live'. A whakapapa was not usually chanted down the horn of a recording phonograph in six-minute chunks. Such recordings can shape understandings of culture and cultural practices, and should be heard with this in mind.¹⁴

The permanence of recordings made them sources of pleasure in their own right. They were objects that could be owned and given as gifts. My record of 'Tip-Toe Through the Tulips' has a postage stamp-sized sticker attached to the label that reads: 'A present from TOT. Jan 22nd. 1930'. 'TOT', whoever he or she was, saw fit to mark the gift of the record by attaching this note (or perhaps it was annotated by the recipient). The good condition of the disc indicates that the recipient and his or her descendants looked after it carefully. Its value was probably sentimental rather than economic. The physical objects on which sound was stored might themselves become important as items to be treasured, as receptacles for emotions and memories above and beyond any that might have been stirred by the music in the grooves. The permanence of recordings as physical objects and the permanence of the music they preserved allowed listeners to make songs their 'own'. A film seen on a first date, a song heard from a radio, a record given as a gift: all of these could be used to mark moments of personal importance.[15] Records could be records of more than just sounds.

Repetition was the third aspect of sound altered by the new technologies. The industrial production of records and films began in the United States and Europe during the 1890s. This gave rise to new industries that involved the production, distribution, marketing, retailing and eventual broadcasting of these products. Mass production of sound led to its mass repetition by records, films and radio broadcasts, all of which transformed ideas about musical listening.

One effect of the replication of musical experience was well captured by the philosopher Theodor Adorno: 'The ability to repeat long-playing records, as well as parts of them, fosters a familiarity which is hardly afforded by the ritual of performance.'[16] The familiarity developed by repeated listening works on several levels.

Recorded sound allowed listeners to know music or performers they may never have heard live, and to know them intimately. Enrico Caruso never toured New Zealand but his recordings were available here from 1903. A 1936 record catalogue listed recordings by then relatively unknown composers such as Clement Jannequin and Claudio Monteverdi, along with many lesser-known pieces by the 'immortal greats' such as Beethoven and Mozart.[17] Alongside this European art music there were thousands of jazz, blues, country, folk and 'ethnic' recordings available. A familiarity with a wide range of music was made possible through the media of recordings that could be repeated until the strangeness and novelty of the sounds wore off.

A more direct sense of Adorno's use of 'repetition' is that listeners could come to know a piece of music in ways precluded by live performance. Repetition of a piece (or part thereof) can lead to a sort of familiarity that removes what Walter Benjamin referred to as 'aura'.[18] The well-worn

> *The joy of recorded repetition was that you could keep trying the Lambeth Walk in the comfort of your own home until you got it right.*
> Evening Post, *12 December 1938, 5*

opening bar of Beethoven's Fifth Symphony is a case in point: 'da-da-da daaaaah' is now so well known that the music retains little of its original revolutionary symbolism or fervour. Contempt is often not far from familiarity. But this repetition can also be a source of comfort. Many people listen to recordings over and over for pleasure.

The effects of the commodification of music and the construction of a music industry based on replication and repetition are endlessly argued. Adorno viewed the effects of mass production and the replication of musical experience as deadening creativity and spontaneity on the part of listeners, composers and performers. By his account, 'the phonograph record is an object of that "daily need" which is the very antithesis of the humane and the artistic, since the latter cannot be repeated and turned on at will but remain tied to their place and time'.[19] Adorno was not alone in his objections to the industrial production of music on moral, political and aesthetic grounds. The band leader John Philip Sousa wrote about 'the menace of mechanical music' in 1906, and the composer Constant Lambert railed against 'the mechanical stimulus' of recorded sound in 1934.[20] Some New Zealanders were also concerned about the loss of musical 'aura'. The ways in which they tried to control recorded sounds illustrated concerns about the roles of high and low culture and the perceived dangers of popular culture.

Some writers argued that the mass production of music was a good thing. The gramophone, the radio and the cinema offered chances to hear and appreciate music no matter where the listener might be. The technologies spread cultural goodness and were useful educational tools as well as desirable machines of pleasure. But mass-produced music was acceptable only so long as it was of the 'right' kind. This usually meant classical music.

Controversies over 'canned' or 'mechanical' music in New Zealand were often debates about high and low culture: that is, mainly British and European high culture versus American popular (or low) culture. Such debates may seem quaintly obsolete in a world of globalised popular culture dominated by transnational entertainment conglomerates, but New Zealanders still discuss and debate the sounds we hear from machines. These machines may be tablets, phones or laptops rather than phonographs, radios or cinema orchestras, but the sounds we hear are portable, repeatable and durable; we are living through the aftermath of a revolution in listening that began with technologies developed during the last decades of the nineteenth century.

Listening is at the heart of this book. It listens in on echoes of life in New Zealand during a key period when daily life was radically transformed by technologies that captured, stored and replayed or transmitted sound. This is not a systematic history of radio companies, record shops, gramophones, movie theatres, or the rest of the institutions and hardware that were part of the sonic revolution of modernity in New Zealand. Rather, the book offers a series of partial and fragmentary soundings of how technology allowed New Zealanders to hear the world and themselves in new and unpredictable ways. It is about the richness and pleasures of New Zealand's historical soundscapes.

Without further ado, it is time to crank the gramophone, or tune the wireless, or open the Jaffa box as the cinema lights dim, and hearken to the richness and variety of listening in New Zealand's past soundscapes.

Part I | RECORDS

CHAPTER 1

EXHIBITING SOUNDS

•••

Only a few hardy souls braved the bad weather to gather at Blenheim's Lyceum Hall on 15 April 1879. They were there to watch a display put on by the entrepreneur and showman C.J.W. Griffiths. As the rain beat on the roof, the audience members listened intently to a recitation of 'Old Mother Hubbard', followed by 'a line or two from a song, barking, crowing, a cooey and other sounds'.[1] Even on a wet night in Blenheim, this was hardly the stuff of grand entertainment, but these sounds were made by a phonograph that Griffiths was touring around the country. A sonic revolution had arrived in New Zealand.

Not all of Griffiths' exhibitions went as smoothly as that in Blenheim. A public display in Wellington at the Athenaeum Hall had some teething troubles. There was great amusement when the recorded words 'Victoria by the grace of God' were heard as 'a dismal wail'.[2] But if technical hitches like this undermined the advertising ballyhoo that described the phonograph as 'science and mystery, the greatest marvel of the age', the machine aroused interest wherever it was displayed.[3]

The startling effects of recorded sound permanently transformed the ways in which New Zealanders heard themselves and the world. This chapter examines the first few years of this aural revolution by exploring the phonograph exhibitions that toured the country during the last two decades of the nineteenth century. Initially presented as scientific marvels, machines such as the phonograph and the gramophone soon became credible sources of musical entertainment and education, in many cases replacing the piano. They became normalised and accepted through a combination of humour and scientific explanation that demystified the marvel of 'talking machines'. The phrase 'talking machines' itself points to an important factor in their acceptance. The revolution in listening brought about by these machines was initially built on a paradoxical emphasis on their imagined and metaphorical humanity. Just as the piano was used to express the human soul of music, the canned music machines came to be heard as serious musical instruments that repaid attentive listening.

New Zealanders heard the world through their phonographs and gramophones. They became familiar with the sounds of internationally famous performers without seeing and hearing them on stage. Local publications such as *New Phonogram* and the *N.Z. Record Herald and Kinetoscope News*, along with the manufacturers' manuals, pamphlets, newspaper articles, magazine stories and advertising, supplied information and gossip about recording artists and new developments of the technology. These international sonic connections have been ignored by local historians who have concentrated on a later period, when a 'national' sound was recorded.[4] But a focus on the early years of recorded sound, and a recognition of its imported nature, offers an opportunity to respond to Peter Gibbons' call for scholars to become less preoccupied with 'New Zealand's place in the world' and to pay more attention to 'the world's place in New Zealand'.[5]

People develop uses for technology that suit their lives, rather than shaping their lives around the innate qualities of any given technology. A prime example is that other acoustic technology of modernity, the telephone. Originally intended as a tool for the commercial world, it rapidly became a source of pleasure and fun as users explored the new possibilities of communication it offered.[6] Unexpected uses were likewise found for the phonograph and gramophone.[7] Like listeners elsewhere, New Zealanders were quick to take up these new acoustic technologies because they could find meaningful and pleasurable uses for them in their lives.

• • •

The phonograph was seen and heard in New Zealand just 18 months after its development by Thomas Edison and a team of technicians at Menlo Park, New Jersey, in 1877. Driven by hand or a clockwork engine, the phonograph stored sounds on strips of tinfoil. The machine was difficult to operate and the recordings could be played back only three or four times before disintegrating. C.J.W. Griffiths, the entertainment promoter and impresario who toured the device throughout New Zealand, followed the pattern of his counterparts in the United States, Great Britain and Australia. His exhibitions were part of a sophisticated international publicity campaign engineered by Edison to draw attention to the latest product of his research facility.[8]

EXHIBITING SOUNDS

Thomas Edison with the machine that changed the way we heard the world: the tinfoil phonograph. For the first time in history sounds could be captured, stored and played back. Library of Congress Prints and Photographs Division, Washington DC, LC-DIG-cwpbh-04044

For those who attended the 1879 exhibitions, the word 'phonograph' and the existence of audio technologies were not entirely surprising. New Zealand ears were primed. 'Phonograph' and its derivatives – 'phonographer', 'phonographist' and 'phonography' – were already used to describe those who followed Isaac Pitman's system of shorthand.[9] The phonautograph, a device developed by L.M. Scott that made images of sound waves on glass covered in carbon blacking, had been described in the New Zealand press.[10] And press accounts of the invention and exhibition of the tinfoil phonograph prior to Griffiths' tour had framed the device as further proof of Edison's genius and the general onward march of science and progress.[11]

The demonstrations began with a brief description of the machine, its invention and its functions.[12] Audience members were then invited to make recordings that were played back to them, and also heard recordings made prior to the show. The phonograph was presented in the context of a lecture rather than as part of a carnival entertainment, which emphasised the serious nature of the technology. But the wonder of a machine that could reproduce sounds – talk and sing – was always emphasised as well. American audiences typically displayed a mixture of enthusiasm and amazement when they first heard Edison's invention, marvelling at the idea of a machine that spoke.[13] New Zealanders were similarly impressed.

One reporter who attended a Wellington exhibition described the playback of his own voice as 'very uncanny' and added that a cornet solo made in Dunedin was 'reproduced with capital effect'.[14] Another commentator remarked on the faithfulness of the recordings of a barking dog, as well as of people laughing, singing and reciting nursery rhymes.[15] In Christchurch, the phonograph was said to be capable of capturing 'every trick of a speaker's voice and the notes of some gifted singer'. This reporter went on to stress that the machine was capable of storing the sounds and reproducing them with 'startling fidelity' either in a moment or some years afterwards for 'the delight of a new generation'.[16] The phonograph was 'an acoustic wonder' and its ability to reproduce sounds aroused 'surprise and astonishment'.[17] Sound had been made permanent, portable and repeatable.

But the seriousness and success of these demonstrations were often undermined by the unreliable nature of the phonograph. In New Zealand and elsewhere, there was much hilarity when at some of the performances the technology failed. The sounds made by a malfunctioning phonograph at one demonstration at the Athenaeum Hall in Wellington caused gales of laughter that disrupted both a prayer meeting and a lecture, leading

> ATHENÆUM HALL
> (FIRST FLOOR).
> FOR SHORT SEASON,
> Commencing
> SATURDAY, 17th MAY.
>
> SCIENCE AND MYSTERY.
>
> The Greatest Marvel of the Day,
> THE PHONOGRAPH!
> Speaking and Singing Machine
> Speaks any language
> Sings any Song
> Plays any Cornet Solo.
> Every Sound that can be uttered by the
> Human Voice
> THE PHONOGRAPH REPEATS!!

A phonograph exhibition aroused a sense of wonder through a combination of 'science and mystery'. This advertisement promises both entertainment and edification. Evening Post, 14 May 1879, 3

to 'much tearing of hair and muttering of curses'.[18] On this occasion, it is likely to have been the noises of the phonograph's listeners rather than the machine itself that led to such acoustic confusion. When the machine malfunctioned, which seems to have happened regularly, the results could also be amusing. At a public exhibition in Wellington, technical glitches led to some recorded lines from *Othello* being described as 'wonderful noises, not at all Shakespearean'.[19] One reporter at a private Wellington show wrote that the sound of the machine was 'of a peculiar character and more like the echo of a voice' and that it repeated words 'in a very drunken manner'.[20] Anthropomorphising the phonograph was a typical response to this wondrous, sometimes bewildering technology. Another was the use of the term 'talking machine'. Humanising the technology made it comprehensible, as did the sober scientific explanations and the humour it provoked when it malfunctioned.

While the science behind the machine was admired, many people initially viewed the phonograph as an entertainment rather than a serious and useful scientific device. There was little or no speculation in the New Zealand newspapers about its potential uses for business, education, scientific research or domestic entertainment. It was a curiosity that embodied the power of contemporary science and technology.

After the 1879 tour, recorded sound seems to have languished in New Zealand, as it did elsewhere. The Australian-based entertainer J. Kohler toured New Zealand with a variety and waxworks show that included an Edison phonograph during 1880–81.[21] A lecturer named C.A. Edwards gave some talks in Wellington on scientific topics that included a demonstration

Douglas Archibald of Oxford was marketed as a sensible man of science whose phonograph exhibitions were sober and serious demonstrations of modern progress and technology. They were also *a lot of fun.* NZ Observer and Free Lance, *7 February 1891, 15*

of a phonograph in 1887.[22] By then Edison had turned his attention to different projects, and other researchers struggled to find a more durable recording medium than tinfoil. Emil Berliner demonstrated the disc-based gramophone in 1888, an event that was noted in New Zealand.[23] Then, in that year, Edison returned to the science of sound with the development of a reliable phonograph he dubbed the 'Perfected Phonograph', along with the durable wax cylinders that marked the dawn of commercial recording.[24] Three years later, Douglas Archibald gave a series of exhibitions that created huge interest and began to transform New Zealand's sonic cultures.

Archibald had demonstrated Edison's Perfected Phonograph in Britain in 1889. He visited Edison in 1890 and bought the latest model before touring the machine through Australia. After astounding audiences there, he sailed for New Zealand and gave exhibitions throughout the country, beginning in Dunedin in December 1890 and continuing during the first half of 1891. He returned to Australia, then sailed for Britain in 1892, giving more exhibitions as he went in Burma, India and Sri Lanka.[25] For him, New Zealand was just another listening post in an international network of phonographic shows. In publicity for the events, Archibald was presented as a serious, trustworthy and sober man of science as opposed to a vaudeville showman. Listeners were promised music from Britain and from Edison's own laboratory, and a demonstration that would electrify them with the 'wonders of modern science'. This was a seriously fun rational recreation.

Archibald's New Zealand tour, under the auspices of the trans-Tasman entertainment promoter William MacMahon, took in most parts of the country, attracted large crowds and was covered extensively in the press. In contrast to Griffiths' exhibitions of 1879, Archibald presented carefully planned and smoothly executed shows that combined not only high seriousness with amusement but also new sonic and visual technologies. Each demonstration began with a lecture accompanied by magic-lantern slides that explained the machine, its development and its potential. Archibald then played a series of cylinders, most of which had been made overseas by internationally famous performers. He also played recordings of New Zealand performers such as Salvation Army bands from Christchurch and Napier, and the Wellington Garrison Band. His final act was to invite audience members and musicians to come onto the stage to make recordings on the spot. These were then played back, much to the audience's delight.[26]

The recordings Archibald brought to New Zealand also included speeches by famous people. One of these, by William Gladstone, 'excited great curiosity, many of those among the audience, born and bred in the colonies, naturally evincing a desire to hear the voice of the great English statesman'.[27] Messages from local dignitaries were recorded too. Robert Stout, lawyer and statesman, was recorded live on the stage and then directly played back to great amusement in Dunedin.[28] Archibald's recordings of former Governor George Grey in Auckland included messages to Edison and to the people

This is allegedly the recording made by George Grey. However, the museum catalogue states that the original record was broken and another substituted, with the same message spoken in a different voice. The cylinder lies on its side. Auckland Museum Tamaki Paenga Hira, 1965.78.20, 356, ocm0103, col.0356

of New Zealand, and were played to great acclaim during the rest of his tour.[29] Contemporary accounts described how audiences listened with 'keen attention' and 'earnest attention', and were 'greatly astonished'.[30] When the Gladstone recording was played in Dunedin, 'a dead silence reigned, then loud applause followed'.[31] Hearing a machine reproduce sound that had been made months before on the other side of the world, or minutes before on the stage, was both a marvel and an enjoyable audio experience in its own right. New Zealanders enjoyed the sounds of their masters' voices.

Audiences were also enthralled and impressed by the clarity and quality of Archibald's recordings. Press commentary was admiring, too, despite some initial scepticism. 'Orpheus', the music and drama critic for the *Sporting Review*, was at first uncertain: 'I am fired with a burning curiosity in respect to the phonograph. We are promised the voices and *ipsissima verba* [the very words] of great orators, singers, and humourists, but will the reality come up to the expectations excited?'[32] After the exhibitions in Auckland, a clearly delighted 'Orpheus' wrote: 'My vocabulary fails to supply epithets to fit the case of that bewildering Phonograph; strange, unique, marvellous, wonderful, are all inadequate – miraculous is perhaps the best word.'[33] Despite the meticulous explanations provided by Archibald in his lectures, the ability of the machine to reproduce sounds remained uncanny and somewhat baffling.

In the advertising for his displays, Archibald's name was always followed by 'MA, Oxon', enhancing both his and the phonograph's credibility. Edison's name was already a byword for genius and a rugged, individualistic 'can do' attitude that embodied the Victorian ideals of self-sufficiency, self-reliance and self-advancement. Now Archibald's smooth, authoritative lectures placed the machine in the mainstream of contemporary scientific development and progress. In his lectures and in interviews, he stressed the business and scientific applications of the device. These included its uses in office dictation and as a form of postal communication.[34] This emphasis on utility was in keeping with the list of potential uses of the phonograph that Edison outlined in the *North American Review* of June 1878. Of the 10 ideas he mentioned, only one, 'Reproduction of music', was concerned with leisure.[35]

For many New Zealand listeners, however, the phonograph displays and the glimpse they offered of the machine's fanciful possibilities were a pleasure in themselves. A cartoon from the *New Zealand Observer*, for example, showed the governor as a human/phonograph hybrid repeating

The anthropomorphised phonograph as sketched by a New Zealand witness of an exhibition by Douglas Archibald. Centre stage, the urbane promoter of the exhibitions, William MacMahon, rides a profitable phonographic steed.

NZ Observer and Free Lance, *14 February 1891, 8*

Justifiable Precaution.

CHOLLY: Sir, I love you daughter fondly and wish to marry her. She told me that you would listen to me—but I did not know you were deaf.'

PAPA: Oh, I ain't. Fact is, I just thought I'd take down your statement in this phonographic recorder, because, you see, Marie is my homeliest daughter and I can't afford to take any chances. She's yours, sir.

You'd think even the dimmest of suitors would have spotted the phonograph lurking on the mantelpiece behind Papa's stern visage. But a man's recorded words were his bond, and marriage contracts were serious affairs in 1898. NZ Observer and Free Lance, *31 December 1898, 15*

the same things over and over. In the top left corner was the British singer John Sims Reeves, famous for the longevity of his career and his never-ending sequence of farewell concerts.[36] Journalists recording a fiery speech were depicted with phonograph horns instead of mouths. A bashful suitor made his marriage proposal by phonograph. Anthropomorphised machines delivered messages to stereotypical wives and mothers-in-law.

Such depictions were designed to allay fears about the phonograph's unnatural ability to reproduce human sounds. They also aligned the phonograph with communication technologies such as the telephone and the telegraph. Edison had envisioned the machine as a substitute for letter writing, and Archibald discussed the establishment of Phonograph Bureaus that would allow people to make recordings that could be posted.[37]

Other New Zealand commentators echoed these light-hearted responses to the machine but also acknowledged some fears. A writer from the *New Zealand Graphic and Ladies' Journal* imagined the machine being miniaturised so it could be concealed and used to entrap suitors and spouses by 'the recording of testimony of the most incriminating kind – the testimony of one's own voice'.[38] What was previously ephemeral – the spoken word – had now become as permanent as the written word, and might even become as binding as a permanent legal document. The article went on to quell those fears: 'Many a phonograph studiously regulated and concealed by an astute mother on her daughter's person will incomprehensibly get deranged and return giving no echo of Lothario's vows, possibly owing to the maiden's complicity or the indiscreet pressure to which it has been subjected'.[39] Another cartoon depended on a similar gag, with the cunning father pretending to be deaf while the all-hearing phonograph captures the young man's words as well as he does. The cruel humour relies on the phonograph as a trap to get a man for the 'homeliest daughter'.

Both fictional and factual accounts of the role of the phonograph in courtship, marriage and divorce were fairly common throughout the western world before World War One. While usually playing on contemporary gender stereotypes, these anecdotes showed that the capacity of the talking machine to capture and store sounds such as the voice had its dangers as well as its pleasures.[40]

The phonograph was also a way of making the dead speak, an ability that highlighted the 'weird', 'uncanny' and 'supernatural' aspects of the storage of sound.[41] Even the living sounded otherworldly. One reviewer noted that the 'speeches by famous men are given in splendid style. It almost seems as a voice speaking from the grave and the enunciation is very nearly perfect.'[42] Another wrote that the machine 'enabled us to embalm the voices of those we love as easily as a photograph enables us to preserve and recall their features'.[43] Edison had listed the idea of a 'Family Record' among the potential uses of the phonograph. This record would include sayings, speeches and even the last words of family members.

Death, its rituals and customs, played a major role in nineteenth-century Western culture, and the early uses of the machine, both real and imagined, were placed within this matrix of mourning and memory.[44] During the 1890s and 1900s accounts of the use of the phonograph to remember the dead appeared in New Zealand as well as overseas.[45] James Joyce was to later mock this in *Ulysses*, inscribing a beautifully onomatopoeic version of the phonograph's re-inscription of the human voice complete with surface noise:

> Besides how could you remember everybody? Eyes, walk, voice. Well, the voice, yes: gramophone. Have a gramophone in every grave or keep it in the house. After dinner on a Sunday. Put on poor old greatgrandfather Kraahraak! Hellohellohello amawgullyglad kraark awfullygladaseeragain hellohello amarawf kopthsh. Remind you of the voice like the photograph reminds you of the face.[46]

Joyce's fictional grave-side auditors may well have listened as intently as the people who heard Archibald's demonstrations.

• • •

A wax cylinder can capture only a small part of the audio spectrum, and constant surface noise, similar to the sound of frying bacon, is an inescapable feature of the technology. To modern ears, even a freshly made wax cylinder can sound muffled and ghostly. Bass and treble responses are very limited, and wax recordings of many instruments, such as guitars and pianos, lose so many overtones that it can be hard to identify what is actually making the sounds. At least, that is how we hear such recordings now. But the use of terms like 'fidelity' by Archibald's auditors indicates the relativity of listening. The phonographically rendered sounds heard by New Zealand audiences in 1891 were amazingly clear and distinct. Many contemporary accounts of Archibald's shows called attention to the clarity of the recordings especially when heard through earphones.[47] These audiences heard the phonograph in a cultural context wherein the reproduction of sound itself was a scientific marvel. Listening was shaped by expectations and situations, just as it is today.

The contexts in which New Zealanders heard recorded sound changed during the 1890s as the new audio technologies became the basis of an international and commercially viable entertainment industry. As the quantity and variety of available commercial recordings increased, and New Zealanders became more familiar with the technology, touring exhibitions of the phonograph began to emphasise entertainment over education.[48]

During 1898, for example, Mr and Mrs H.A. Salmon displayed a phonograph in Wellington. The main attraction of this show seems to have been the machine's volume, and the large quantity and variety of recordings played. Exhibitors no long needed to make their own recordings: the Salmons, for example, had more than 130 records, including 'The Lost Chord', 'The Anvil Song' and 'The Laughing Song'.[49] (The Salmons' phonograph was also used as part of a vaudeville show that featured singers, dancers and comedians.[50]) A report of a Wellington show in 1899 described the phonograph as 'a familiar means of entertainment nowadays'.[51]

The New Zealand exhibitions of the late 1890s point to the development of a commercial entertainment culture based on recorded sounds. The records played were not made in New Zealand; they were manufactured and distributed internationally by European and American-based companies. New Zealand became just another market as the new music industry of recorded sound expanded into a global economy. That enterprise brought with it major changes to New Zealand's music retail trade during the early years of the twentieth century. Records, phonographs and gramophones began to encroach on the sales of sheet music and musical instruments, particularly pianos. The sounds for exhibition now became sounds for sale.

Chapter 2

SELLING SOUNDS

•••

New Zealanders were familiar with recorded sound by the end of the nineteenth century. They had read about it in newspapers and magazines. They had seen photographs, drawings and cartoons showing the phonograph and its myriad possibilities. Many had seen and heard the machines in action during the exhibitions by Douglas Archibald and others. As thoroughly modern people in touch with the latest international trends in technology and fashion, New Zealanders were fascinated by recorded sound and wanted to experience it for themselves. But how?

Record stores did not exist, so individuals were obliged to import cylinders, discs and the machines that played them at some considerable cost and inconvenience. The machines, too, were often bulky and did not fit well with domestic decor. It was also not clear what the machines were for. Were they serious instruments of science, as Archibald Douglas had presented them? Or were they sonic fun machines that were simply the fad of the moment? Did they belong in the parlour? How were they to be heard? New technologies raise new anxieties as well as possibilities, and the phonograph was no exception. Fortunately for these early adapters, the nascent international recording industry was on hand to supply guidance and allay apprehensions.

•••

The business of recorded sound in New Zealand began in 1900 with the arrival in the country of Peter H. Bohanna, a representative of the British-based Gramophone Company. Charged with establishing markets in Australasia and the Pacific, Bohanna travelled the country establishing contacts and contracts with various firms that distributed the company's machines and recordings.[1] These included shops such as Charles Begg & Sons, one of New Zealand's most prominent chains of music outlets, and the Dresden Piano Company, along with general stores such as Dennes Bros and importers such as Hyams. Bohanna handled the distribution of records and machines to these businesses from his office in Pitt Street, Sydney. He was in turn answerable to the company's head office in London. This arrangement was typical of the way the early recording companies divided up and managed the global markets for recorded sound. In 1901 the British Gramophone Company came to an agreement with Eldridge Johnson's Victor Company: the Gramophone Company sold its products in Europe, Russia, Japan and the British Empire, while Victor would sell everywhere else.[2] Other companies also pursued worldwide markets at the same time.[3] When it came to recorded sound, New Zealand was a branch office of a global industry – and a relatively minor office at that.

Bohanna sold his products through pre-existing retailers. The first New Zealand retail spaces dedicated to recorded sound began in 1901 with Thomas Holton and his sons, who established a chain of stores throughout New Zealand under the name 'The Talkeries'. By 1914 the franchise had shops in Auckland, New Plymouth, Masterton, Wellington, Christchurch and Dunedin.[4] Though focused exclusively on recorded sound, these stores were to some degree modelled on music shops such as Begg's and the assorted piano companies. Customers of The Talkeries could purchase machines, spare parts and records; they were also offered the chance to hear as well as make recordings. Demonstrations of the equipment were carried out in the same way as musical instruments were demonstrated in music stores.

In a photograph of the interior of a Talkeries shop in the Wairarapa township of Masterton from about 1909, we see that the cylinders are tidily displayed on the front counter and on shelves lining the walls. Catalogues of new recordings are conveniently at hand on the counter, alongside the latest machines. Posters on the wall advertise recording stars and recordings from the Pathé label. The notice above the office door states that records can be heard for payment of 2d, but those by the opera singers Enrico Caruso, Ada Crossley and Nellie Melba will cost 6d. This fee was no doubt intended to deter non-buyers from enjoying free entertainment. But the higher charges for playing opera records also reflected the role of opera as high culture: it was more valuable than popular music. And the technology's own social and cultural status could only be enhanced by its association with high art.

Seriousness and respectability were recording industry ideals in the years before 1914. Making retail outlets like The Talkeries shop in Masterton

The Talkeries chain spanned the country and supplied everything a music lover needed in tidy, respectable shops that enhanced the prestige of recorded sound.
Interior of 'The Talkeries', Alexander Turnbull Library, Wellington, Ref: 1/2-043062-F

into respectable music shops was part of this drive.[5] The recording of internationally famous singers and the marketing of expensively priced, exclusive record labels were also important in elevating recorded sound from a carnival attraction to a serious art that could be safely enjoyed by the middle classes. Advertising for records and machines often depicted formally attired people in plush settings enjoying the sounds of recorded opera.[6]

But the Masterton Talkeries and other shops catered to wider tastes than those of a 'highbrow' persuasion who liked their opera. The lists of new arrivals regularly advertised by The Talkeries in local newspapers make it clear that popular music was a big seller too. A list from February 1909, for example, features the Scottish singer Harry Lauder. Ada Jones, one of the first women singers to record, appears in two duets with Len Spencer and Bill Murray. Both of these men were huge stars in their own right on record and stage. Selections from the American comedian Cal Stewart were further instalments in his famous comic monologues about the adventures of Uncle Josh and the rural japes and mishaps at 'Punkin [sic] Centre'.[7] Instrumental selections by bands and soloists round out the selection.

The hit list just in time for Christmas 1908. These are just some of the recordings that could be bought from Masterton's Talkeries. Record stores became social spaces where people could while away time chatting and listening to new sounds. Wairarapa Age, 27 February 1909, 3

```
EDISON RECORDS for DEC., 1908
          (TWO-MINUTE)
10008  Christ is Come        Edison Concert Band
10009  Always Me             Byron G. Harlan
10010  Taffy                 Ada Jones
10011  Petite Mignon (Oboe)  Caesar Addimando
10012  When Darling Bess First Whispered
         Yes                 Manuel Romain
10013  My Brudda Sylvest     Collins and Harlan
10014  Everybody Knows It's There
                             Edward M. Favor
10015  Fun in a Barber Shop  Vess L. Ossman
10016  Uncle Josh's Arrival in New
         York City           Cal Stewart
10017  The Widow Dooley
                             Ada Jones and Len Spencer
10018  I'm Glad I'm Married  Billy Murray
10019  In Lover's Lane       Edison Concert Band
10020  The Sons of Uncle Sam Edward Meeker
10021  Last Day of School at Pumpkin
         Centre              Cal Stewart
10022  My Rosy Rambler  Billy Murray & Chorus
10023  Kentucky Patrol
                    American Symphony Orchestra
10024  Yours is Not the Only Aching
         Heart              James F. Harrison
10025  Oh, You Coon  Ada Jones and Billy Murray
10026  What You Goin' to Tell Old
         St. Peter?          Arthur Collins
10027  Song of the Mermaids
                    Venetian Instrumental Trio
10028  I Don't Want the Morning to
         Come               Frederic Ross
10029  So Do I        Knickerbocker Quartette
10030  Christmas Morning at Clancy's
                             Steve Porter
10031  Uncle Sam's Postman March
                             Edison Military Band

       "The Talkeries,"
     BANNISTER ST., MASTERTON.
           P.O. Box 21.
```

Like other record shops, The Talkeries also utilised marketing opportunities. Marie Narelle was an Australian singer acclaimed internationally for her performances of Irish and Scottish songs, although she also sang some opera.[8] She toured New Zealand in 1905, and the Masterton Talkeries was quick to cash in. The shop ran a series of advertisements announcing available records by Narelle. There was even a special display in the shop window, which was dressed in green to honour the singer's Irish roots and music. A local newspaper pointed out that this was 'usual in other countries but it was a novelty in Masterton'.[9]

Recitals of recorded sound were another technique used by record sellers to make audio technology respectable, 'musical' and popular. These were modelled on the recitals held in music shops to demonstrate instruments. The concert programmes typically contained a mixture of light and serious musical items that also showed off the capabilities of the machinery. At the Wellington Talkeries in 1905, for example, an 'All Stars Artists' concert featured a variety of music and recitations by performers such as The Edison Concert Band, B.G. Harlan, Ada Jones and Edgar L. Davenport. Admission was free.[10]

37

Recitals of this kind were usually held on retail premises, but some were in more public locations. A free recital at the Wellington Town Hall in 1909 was given to show off the Auxetophone, a phonograph variant that used compressed air to generate loud volume.[11] The event's advertising emphasised the star qualities of the featured singers: Nellie Melba, Caruso and Antonio Scotti.[12] The subsequent review discussed the technical aspects and origins of the device, and commended the quality of the sound reproduction by describing it as the 'perfection of the newest Gramaphone [sic]'.[13]

Adding to the technology's credibility, the phonograph recitals sometimes blurred the boundaries between recordings and human presence. The Wellington store George & Kersley advertised two recitals in 1908.[14] Each was billed as a 'grand vocal and instrumental concert', but there was no mention of recordings. Instead, the advertising copy stated that 'Madam Melba will sing *Ava Maria*, Miss Amy Castles will sing *La Serenata*' and so on through a list of world-famous and popular performers including Caruso, John Harrison, Harry Lauder and Dan Athenaeum. No one would have been fooled into thinking these artists would be present. But the implication was that the mechanical reproduction of their performances was as good as the real thing. There was a serious side to this playful claim that musical simulacra were equivalent to real musical events. The technology was respectable and worthy of being taken seriously in musical and artistic terms, and it should be listened to in the same way as human musicians were.

The New Zealand shops that catered to human musicians followed overseas trends by adding the new audio technologies to their stock. Charles Begg & Sons opened a gramophone and phonograph department at their Dunedin shop in 1906. By 1911 their stores in Wellington, Timaru, Invercargill and Oamaru also had such departments.[15] This expansion was driven by demand. In Dunedin a Begg's newsletter described phonographs as 'instruments', and noted that the company's recording retail department had expanded dramatically in just five years.[16] The use of the word 'instruments' here is important. Proponents of the new technology often described their machines as musical instruments. This was part of their strategy to gain credibility, acceptance and cultural capital. Opponents, on the other hand, often based their criticism of the technology on the idea that the machines took away the human element of music.

An influential statement of the opposition to recorded sound appeared in America in 1906, and was noted in New Zealand. 'The Menace of Mechanical Music' was written by the famous band leader John Philip Sousa. After describing recordings as 'canned music', Sousa attacked the spread of the phonograph as inimical to musical ability, education, composition, childhood development and military prowess. He described music as teaching 'all that is beautiful in the world', and asked that it not be hampered 'with a machine that tells the story day by day, without variation, without soul, barren of the joy, the passion, the ardour that is the inheritance of man alone'.[17] He then turned his attention to the unfairness of contemporary copyright legislation that deprived people such as himself of the ability to profit from recorded music.[18] This issue tended to be overlooked by many who agreed with his thunderous denunciation of the phonograph itself, though others argued for the educational and cultural benefits of recorded music.[19] In New Zealand the *Otago Witness* theatre critic 'Pasquin' concurred with Sousa's sentiments, and added that the phonograph's popularity had caused a slump in local sales of musical instruments such as guitars, mandolins and pianos.[20]

For those selling the technologies of mechanical music, the sort of opposition expressed by Sousa was overcome by emphasising the refined nature of the machinery and its advantages over normal musical instruments. A Begg & Sons advertisement from 1909 for the Gramophone Grand machine is typical of this approach:

> This superb instrument is not a phonograph as it is generally called. It is a piece of workmanship calling for the highest skill that science can give. It has ceased to be a musical instrument. It is the actual art, the living soul, of the greatest singers, ready at all times, to wake into melody at a touch.[21]

Here the machine is presented as a precisely made piece of modern technology that is also animate and human. In this account, the reliability of the machine that Sousa derided as 'without soul, without variation' gives it an advantage over musical instruments, as it captures (or cans) 'the living soul' of great artists and makes it available at any time, to anyone, anywhere, even in New Zealand. Spontaneity and inspiration, which Sousa argued were the marks of 'true art' along with human presence, are here ignored in favour of instant gratification and repetition. At the same time, technology is portrayed as an authentic conduit for human presence or the living souls of performers, just as the instruments and voices of musicians and singers are tools of expression. Contra Sousa, to listen to the phonograph is as authentic an experience as hearing the singers perform live.

The alliance between the new technologies of sound and traditional music shops worked to their mutual advantage. The new technologies gained cultural capital and credibility by being associated with 'the actual art' of great musicians. The music shops tapped into a burgeoning market and lucrative opportunities as audio technologies moved from the public stage to the private parlour.

∴

During the years from 1900 to 1914, the makers and sellers of the new audio technologies concentrated their efforts on getting phonographs and gramophones into private households. Andre Millard wrote of the phonograph in America: 'The important selling points were cultural rather than technological.'[22] That is, selling the new audio technologies on their technological sophistication was not going to be enough to compete with the cultural status the piano had as an expressive instrument. At this time, the piano dominated music-making in homes.[23] Technological developments during the nineteenth century had allowed mass production of the piano's cast-iron frames and complicated mechanism, or action, putting the instrument within reach of all but the poorest members of society. The piano had thus become a mass-produced machine as much as any phonograph, but its complex social, aesthetic and cultural roles, as well as its ubiquity, gave it a dominant position in the world of music.[24]

Some of these points were highlighted in a poem that appeared in 1910 in the *New Phonogram*, which was published in Sydney and distributed in New Zealand through phonograph and music shops. Its contents included accounts of new recordings and anecdotes from Europe, America, Australia and New Zealand that highlighted the uses and pleasures of the phonograph:

Shut That Old Piano Up (written for *The New Phonogram*)

Oh, shut that old piano up, let me have peace awhile;
It only makes my poor head ache, its ev'ry note is vile.
There's not a chord of it in tune, in pity let it be;
Such dreadful discord will bring on a fit of lunacy.

How can you think an ear refin'd can listen to such trash?
Had I my way, this day, the horrid thing I'd smash;
No wonder I stop out at night, from such noise to be free,
Your rough attempts at 'Home Sweet Home' makes this no home
    for me.

> If you would music really have, now something fine and grand,
> Songs of beauty, pure and sweet, a military band,
> A brilliant piece of vaudeville, a sketch to make you laugh,
> For goodness sake, here, take my purse, and buy a phonograph.[25]

Changes and refinements in the design and appearance of the new audio technologies were also positive factors in their acceptance as desirable domestic items.[26] By the early 1900s the piano was a discreet, almost cabinet-like object in natural woods. Its narrow depth allowed it to be placed against a wall without too much loss of living space, and its flat top was useful for displaying photographs and ornaments. The piano's mechanism was entirely concealed. By comparison, the early phonograph was unmistakeably a machine and, until the advent of hornless cabinets in 1906, relatively obtrusive. The machine's mechanism was often visible and the sound-reproducing horns were inconvenient even on the smaller models. Manufacturers tried to make the devices more attractive by painting or lacquering the horns and (under the influence of Art Nouveau) adding petal shapes to them.[27] Even so, according to Compton McKenzie, founder of *Gramophone*, the horn offended 'against gentility' and was ugly, cheap, blatant and 'overpowering in a room'.[28] H.L. Wilson, compiler of a 1926 guide to good recordings, reflected that the horned gramophone had been an ornament reminiscent of the Victorian age that many took pains to hide, and that 'more often than not it was relegated to some far corner where its protuberances might not offend'.[29]

The advent of the Victor Company's 'Victrola' machine in 1906 swept aside these objections by using an internal horn. Victrolas, and others that followed, looked like decorative wooden cabinets, somewhat akin to the sewing machines that had already made their way into domestic spaces.[30] The mechanism was hidden until the lid was raised, and even then only the turntable and tone arm were visible. Discreet and designed to fit the domestic interiors of the day, these machines marked a major turning point in the incorporation of the new technology into the home.

With the development of table-top, portable machines during the 1910s the gramophone became even more discreet. The attractiveness of these machines became a selling point. Local advertising for the Mignon Hornless Gramophone claimed that 'apart from its compactness and neat appearance, it is a beautiful piece of furniture, finished in light-coloured oak'.[31] Buying a gramophone now involved more than just sound. The machines were also investments in home decor and status.

THE WORLD'S DIN

These images illustrate the domestication of recorded sound. The phonograph shown at the top is a bulky and industrial-looking machine, whereas the cabinet version below would have fitted elegantly and discreetly into a parlour or similar domestic space.

Top: Frederick Nelson Jones, Alexander Turnbull Library, Wellington, Ref: 1/2-026519-F
Left: Edison phonograph, Alexander Turnbull Library, Wellington, Ref: 1/1-017683-G

These investments could be large. Gramophones and phonographs were sold at a range of prices: in New Zealand the lower-end models cost around £3–4, and more expensive models could be near £40 and even up to £60. These prices remained fairly consistent until the 1920s.[32] The average annual wage for a schoolteacher in 1900 was £120 4s 3d, climbing to £340 by 1920 (for secondary school teachers), so a phonograph or gramophone represented a significant purchase.[33]

Pianos could also be expensive: more exclusive brands could cost up to several hundred pounds.[34] By 1914, however, a budget-price piano from Begg's could be purchased for £20, and hire-purchase systems and second-hand trading made pianos affordable for many.[35] Cultural, rather than financial, capital was the main difference between the piano and the phonograph until the 1920s. The piano required a great deal of investment in time and skills and was a symbol of respectability. It was seen as a serious musical instrument while the phonograph and gramophone were initially regarded as toys. They merely reproduced music rather than made it.

...

The expansion of the available recorded repertoire and the ease with which audio technologies could be operated when compared with the piano were important factors in gramophones and phonographs becoming common household items. The recording industry honed in on these points when selling the new technologies as serious musical listening devices. Despite some initial mystification, New Zealanders became willing consumers, and were all ears to the global recording industry as it got underway during the 1890s. Recorded sound migrated from theatres and exhibitions into the household, and this changed listening forever.

Chapter 3

DOMESTICATING SOUNDS

•••

Having bought a phonograph and installed it in the parlour, the purchaser was faced with a question: what did you do with it? In one sense this was a trivial question. The machine was to be listened to. But how was this listening to occur? Listening to music at home had always involved someone playing an instrument. Codes of behaviour, attitude and etiquette had long been established around this. The new machines required new codes, and manufacturers and retailers were quick to instruct people how they might, or should, use the devices. These publicity materials – magazines, advertisements and manuals – contained guidelines for listening that were both implicit and explicit. Numerous illustrations provided models for practice and aspiration.

A typical example was featured on the cover of the *N.Z. Record Herald* of June 1914. This picture presented a view of middle-class home life that reinforced contemporary gender and social roles. The mother is shown with needlework while the father sits at ease with his pipe, no doubt relaxing after a day at the office. The daughters have their books, while the young man and the boy of the household busy themselves with operating the machinery. Domestic piano-playing and music-making in general had traditionally been a female preserve while male performers dominated the public arenas of concerts and recitals.[1] But technology was part of the masculine domain and so the manipulation of the new music machines was seen to be appropriate for men. This technology was gendered from its early stages, at least in advertising.[2]

The *N.Z. Record Herald* drawing also raised some interesting questions about how to listen to machines rather than to people. The presence of a human performer traditionally provided a focus for attention during musical performances. But if there was nobody physically making the music in front of them, what should an audience focus on while they listened? In

The phonograph tamed and domesticated sound for home listening. Entitled 'A Concert at Home', this illustration from a New Zealand trade magazine offers a miniature 'school for listening' for new phonograph users. NZ Record Herald *70, June 1914*

'A Concert at Home', the conventions of public concert-going were shown as being at least partially upheld. The father's attention is focused on little else but the music.

Attention and raptness were hallmarks of a sensitive audience member, and signified the ability of the hearer to understand, appreciate and enjoy music. This would be appropriate at a live performance both in public and sometimes during formal occasions at home. The mother, with her needlework, and the daughter with her book, are not acting as they would at a public performance. However, they might well carry on such activities if someone was playing the piano informally in their parlour. The young man operating the machine, and the boy watching him, seem to be listening intently to the machine rather than to the music, making sure each disc is operated correctly. No one appears to be talking or making undue noise. In this they mimic the correct behaviour expected at live performances. However, listening in real life would have been far less decorous than the industry's advertising suggested.

The ability of the new music technologies to reproduce voices and instruments meant that fidelity or life-likeness became important criteria in the ranking and cultural capital of the devices, as the poem 'Shut Up the Piano' also implied. Phonographs were better than musical instruments because they reproduced all sounds so faithfully – or at least these were the claims made for them. The listener was invited to listen to the sounds the machine made not as sounds *qua* music or speech, but as gauges of the quality of the machine itself. The sounds of other modern technologies such as industrial machinery, automobiles and engines of all kinds were heard as noise, not music. But the sound of the music machines was a signifier of quality and culture.

The recordings themselves were also objects of acoustic attention. Consumers were subjected to a cacophony of competing claims and hyperbole as manufacturers and retailers fought for market share. Until the demise of the cylinder in the 1920s, the battles between the disc and the cylinder as acoustic media were fierce. The cylinder offered the ability to make recordings but was not easy to use and was more fragile than a shellac disc. The disc was convenient and more robust but could not be used for home recording. Duration and fidelity were important criteria in this rivalry.

A typical advertisement for the Edison Company's Amberol cylinders extolled their duration, as well as their fidelity to both the sound of the recorded artists and the music itself, by claiming the records as the longest made and that 'the famous arias are rendered as originally composed, without hurrying, without omission or slighting of the orchestration'.[3] The listener was promised the entire musical experience, rather than an abbreviated version, thanks to the superior technological nature of the Amberol disc. The pleasure of owning this recording lay not just in its musical qualities but also in the superior technology that delivered a faithful reproduction of an aria. Listeners were invited to enjoy the technology itself. For some this sort of listening evolved into the pursuit of expensive, high-end audio equipment that promised the Grail of total audio fidelity. Dubbed audiophiles, these mainly male listeners tended to pay more attention to the effect of their speakers, turntables or cabling than to the actual sounds of the recordings.[4] But most people were happy enough enjoying the sounds of internationally famous performers in their homes, and the record companies continually harped on this theme.[5]

The artists New Zealanders heard in their homes through the new audio technologies were internationally famous. According to one local advertising writer in 1911:

> No matter how remote your habitation, it [the gramophone] brings within the family circle the actual voices of orators, singers, comedians, and story tellers, who perhaps at that very moment are delighting Metropolitan audiences with the same eloquence, melody, humour and dialect that is coming from a gramophone in the quiet of a country home hundreds of miles away.[6]

The phonograph was a veritable cornucopia of home entertainment with its endlessly varied repertoire to brighten the dullest home. Or so the advertisers would have you believe. Wairarapa Age, *23 January 1907, 3*

The idea of the phonograph connecting people was a common theme. Advertising imagery and words made much of the phonograph's role in family life. Listening to music at home kept young people happy and off the streets. The phonograph made the home 'bright and attractive' and cut out the need to 'go to clubs, theatres, or other places of amusement'. This saved money, yet allowed country dwellers to 'keep in touch with the world of entertainment and music'. But wait, there's more. The phonograph could even 'create a business of spirit enterprise' in young boys (always boys, never girls) through arranging phonograph concerts.[7]

New Zealand listeners were not deaf to these versatile and attractive new technologies. G.E. Lockington recalled that in Paeroa during the 1910s the gramophone allowed 'the hearing of music performed by world artists, with the resultant benefits to keen music lovers'.[8] In 1907 a young Katherine Mansfield wrote delightedly to a friend in England about a 'musical feast', including music by Beethoven, as well as political speeches, that she had enjoyed listening to on a gramophone at her Wellington home. The experience had made her '11,500 miles nearer you than I feel I am'.[9]

• • •

While listening to recordings was an increasingly important part of home leisure, there were also more serious uses for the technology. Much was made of the educational potential of the machines. An article from 1911 reminded New Zealanders that 'the Edison phonograph is more than an ever ready and willing entertainer at the hearth-stone; it is an instructor and educator in the home'. The instruction described here was in music. The phonograph could train the ear and 'teach the child to be critical'. This was to be an education in the accepted canon of musical classics – 'the compositions of the world's greatest masters' – and so reinforced prevailing cultural norms.[10] The gramophone introduced into the home the sort of disciplined, silent and concentrated listening required at concerts, schools and conservatories. Even those who could not read music or play such demanding music could now enjoy the music of the 'greatest masters'. Record companies published guides on music appreciation to accompany the recordings they issued.[11] The disciplined listening of the conservatory and the associated inculcation of accepted artistic hierarchies was now brought into the households of those who could not read a score or play such demanding music.

But not all the uses of the phonograph involved such furrowed brows and studious devotion to the classics. There were plenty of laughs to be had both at and from the machines. The jokes, stories and anecdotes that grew around the uses, abuses and joys of the new technologies further familiarised machines that possessed uncanny and haunting abilities to capture, store and repeat sounds from around the world.

The phonograph had been making people laugh since its first appearance in New Zealand. A machine that could turn people's voices into very weird sounds was bound to be a hit. There was also a large and growing selection of funny recitations, jokes and comic songs available on record.[12] As people became accustomed to the technology, writers and satirists used it more and more to score points or raise a laugh. In a 1910 novel, the comic writer P.G. Wodehouse described a landlady's voice as being 'modelled on the gramophone'.[13] This was not one of his better jokes but it showed that he expected his readers to be familiar with the ways in which audio technologies collected, stored and played back sounds.

Premier William Massey knows what he wants to hear from the new Speaker of the House. In the event it was John Millar, the bowler-hatted phonograph 'Millarphone' (second from the left), who was chosen. Sir George Grey Special Collection, Auckland City Libraries, AWNS-19130619-5-3

In a New Zealand cartoon from 1913, Premier William Massey is shown in the process of selecting a new Speaker of the House following the death of Arthur Guinness. Like any prospective phonograph owner in a shop, Massey listens intently to each candidate. The humour lies in a play on the word 'speaker'. A further implication is that just as a phonograph can endlessly repeat the sounds its operator selects, the Speaker of the House will say what the premier wants. This was a familiar joke: similar cartoons depicted politicians and other important people as machines that noisily repeated their messages over and over.[14]

The image of the phonograph was also used for more gentle humorous effect. In 1911 the New Zealand firm Edmonds promoted the reliability of its baking powder with a pun on its 'good record'. Just as a recording always gave the same sounds to its listeners, this product could be relied on for 'making your baking a success'. Such visual conceits depended on viewers understanding how audio technologies worked. To understand the Edmonds advertisement was to be part of the modern world.

Many of the jokes that circulated about gramophones and phonographs were from overseas sources and publications, beginning even in the early years of the technologies. A typical example was printed in January 1879, before the phonograph was heard in New Zealand: 'Scandalous remark at the Cincinnati Breakfast Table – The phonograph will probably be called a "she" because it repeats everything.'[15] By 1888 the trope of the phonograph as a nagging wife had found its way to New Zealand: 'Edison Agent: Wouldn't you like to buy a phonograph? It will store up everything you say and repeat it to you. Auckland man: 'No, got a wife.'[16] Other typical anecdotes included a story printed in a Wellington gossip column in 1902: 'That a local small boy, of scientific proclivities, has been offered a fabulous sum for a phonograph record of a conversation between his big sister and said sister's idiotic young man.'[17] Another involved a Wellington woman convincing her husband that he snored excessively after capturing him on a wax cylinder.[18] The audio technologies seem to have become part of everyday life surprisingly quickly, and these jokes, stories and cartoons were all important parts of the process.

Some anecdotes portrayed phonograph users taking advantage of the new technologies for their own pleasures. One, from *The New Phonogram*, described how two young women spent several hours trying out the latest records in a store before they 'took their one record and departed while the poor salesman who had made a 2s. sale collapsed on a piano stool'.[19] No location for this possibly apocryphal story was offered, and we know that The Talkeries had a sign that outlined the charges to be paid for listening to records so as to avoid such situations. The story could just as well have been told about people trying on clothes for the fun of it.

● ● ●

The supplies of sounds from around the world were interrupted during World War One, but there was still a great demand for phonographs and gramophones.[20] In New Zealand it became a patriotic gesture to supply records and players as gifts for the troops, along with the usual socks, cigarettes and cakes.[21] Soldiers used recorded music to stave off boredom, as entertainment and for general morale-building. The machines were heard in camps, transport ships and hospitals as well as the trenches.[22] They were highly sought after. A comment from the first issue of the army magazine *Chronicles of the NZEF* spelled this out, and encouraged others to provide

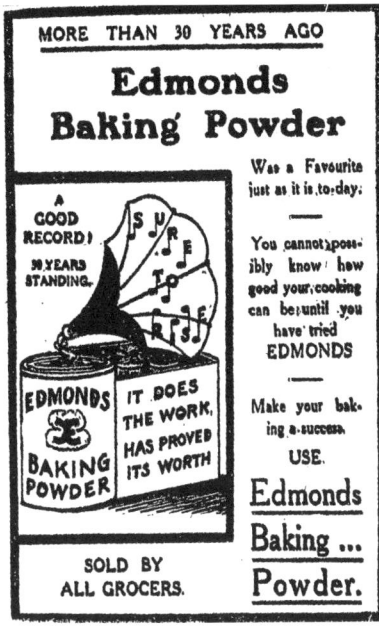

Above: The New Zealand food company Edmonds used a groan-worthy pun on 'record' to emphasise its longevity and tradition.
Evening Post, *2 September 1911, 11*

Right: Gramophones and records went with armies all over the world and helped to while away the hours in camps, ships, hospitals and trenches.
Wanganui Chronicle, *18 February 1918, 4*

such comforts: 'Lieut. and Mrs Dan Riddiford have been so good as to give a certain sum for the purpose of supplying the men in France with Gramophones. They are warmly welcomed by the fellows and go a long way to make cheerful a dismal dug-out or a mouldy billet. Already fifty have been sent over.'[23] The point was not lost on New Zealand's record dealers. Much of their advertising during the war stresses the value of gramophones and records for the troops abroad. Buying records was patriotic as well as fun.

On the other hand, the New Zealand conscientious objector Archibald Baxter noted a more sinister aspect of the gramophone. He recalled a

A group of soldiers on a troopship relieving the monotony of the journey with some music. The ship was the **Port Lyttelton** *and carried men of the 20th Reinforcements, who sailed on 7 December 1916, bound for Passchendaele. Masterton District Library, Wairarapa Archive, Ralph Hopkins papers, 05-118/1-1-5*

recording of 'Onward Christian Soldiers' being jeeringly played on a troop transport as he was publicly humiliated and punished for his opposition to the war.[24] Recorded sound could be a weapon as well as a consolation.

• • •

By the early 1920s, with the war over and the importation of records and players resumed, gramophones and phonographs were well established throughout the country. They were heard in homes, local halls, nightclubs, shops and theatres. Through recordings and the print publications associated with them, New Zealanders were listening avidly to the sounds of an international music industry.[25] Music had been separated from the bodily presence of musicians. The technology had gone from being something of an amusing diversion, as it was often thought of during Griffiths' 1879 tour of the phonograph, to a serious instrument that warranted careful attention. Phonographs and gramophones were pleasurable consumer items in their own right and made sound, hitherto the most ephemeral of experiences, into a mass-produced industry consumed on a global scale. Recorded sounds had become part of the background noise of life and were being put to more serious uses such as teaching.

Chapter 4

TEACHING SOUNDS

• • •

Olive Boyd, a teenager in the coal-mining settlement of Rotowaro near Huntly during the late 1920s, had an illicit pleasure:

> We were banned sometimes from using the gramophone. We weren't supposed to have the gramophone going. So I used to shut the door. I'd have a playing card, you know, an ordinary card and I'd have the machine going. The record would be going around and I'd hold it right in the grooves. I could hear but they couldn't. I'd play the same record over and over.[1]

Boyd's private and clandestine listening with a playing card was an ingenious application of the gramophone's technology, and illustrates the ways in which people turned it to their own ends.[2] There were many ways of listening to the gramophone and Boyd's was one that subverted parental strictures on how the machine was to be used. Schools, record companies and teachers offered similar prescriptions, and these too could be ignored or turned to the listeners' own ends. But this example of sonic subversion illustrates the multitude of listening styles the gramophone allowed. It was heard privately by individuals, but also at gatherings ranging from a few friends in a living room to a packed town hall. The gramophone straddled the divide between the public and the private. It was not as specialised as other audio technologies such as the wireless (heard at home) or the talkies (heard in cinemas). All these technologies were used for education, information and entertainment depending on the occasion. But the gramophone's portability made it a particularly useful educational tool. Many New Zealanders heeded the didactic gramophone between the wars with varying degrees of enthusiasm.

• • •

At school, Olive Boyd may have derived less pleasure from the way her teachers used the gramophone than she did from her secret listening at

HMV's popular book Opera at Home *summarised opera plots and was illustrated with photos of singers in costume. It also listed available HMV opera recordings for home listening.* Otago Daily Times, 13 November 1925, 3

home. Edison had drawn attention to the didactic possibilities of the phonograph as soon as it was developed.[3] Others followed his call as the technology was refined and the available repertoire expanded. By the late 1920s individuals such as the English journalist and record enthusiast Percy Scholes were producing books and articles that highlighted the value of the gramophone for musical education.[4] Local writers recommended books by Scholes in schools and at home.[5] Record companies, quick to climb on this bandwagon, issued pamphlets, books and record sets aimed at schools and home listeners.[6] Educationists such as the English music teacher William Johnson provided detailed and wide-ranging suggestions on how to use the gramophone in the classroom, along with lists of appropriate recordings, from 'Marches for assembly and dismissal' to 'Talks on art, literature, science, classics, etc'.[7] Johnson's list was impressive but the bulk of his book, *The Gramophone in Education*, was concerned with the study of music in terms of performance and, especially, its appreciation.

The idea of musical appreciation was in many ways the ultimate goal of formal musical education and was a site for ideological struggles about aesthetics, pleasure, discipline and music itself during the 1930s. It was important that the 'correct' music was heard and that the 'correct' attitudes to it were inculcated in the minds of its listeners. This involved 'correct'

> **IONA PRESBYTERIAN COLLEGE for GIRLS**
>
> The work of this College is being re-organised and in February, 1925, two courses will be commenced:—
> 1. FOUR YEARS' COURSE OF STUDY FOR HOME LIFE. In addition to the usual class subjects of the school curriculum, this course will include Cookery (Theory and Practical), Plain Sewing and Knitting, Dressmaking, Hygiene, Household Management, First Aid, Home Nursing and Invalid Cookery, etc.
> 2. FOUR YEARS' COURSE LEADING TO MATRICULATION. This course will include English, History, Geography, Arithmetic, French, Mathematics, Science, etc.
>
> Post Matriculation work can also be given.
>
> A course of instruction in Musical Appreciation by means of the Gramophone will be given to all girls.
>
> A wireless set has recently been installed.
>
> Pupils are received from 8 years of age. Prospectus on application to the Principal, who is now enrolling pupils for 1925.
>
> **HAVELOCK NORTH, HAWKES BAY**

Opened in 1914, Havelock North's Iona College was originally a private boarding school for girls from rural families. Music appreciation was an important part of education and Iona boasted the latest technologies, gramophone and wireless.
The Ladies' Mirror, 1 May 1925, 21

deportment and 'correct' aesthetic hierarchies in which art music was valued above popular music and was heard in a disciplined way.[8] The school music syllabus was designed to discipline musical tastes. It also disciplined the ways in which music was ideally heard: the body was to be still and the pupil's attention focused on the music. The concert-hall ideal was to be learned in the classroom via the gramophone.

New Zealand children were taught to listen 'properly' thanks in large part to the work of Douglas Tayler, who was appointed New Zealand's first Supervisor of School Music in 1926. A British-born conductor, composer and organist, he held the post until 1931; he died in California the following year. During his tenure he invigorated musical education in New Zealand through a stream of articles for the *Education Gazette* and other periodicals, along with the publication of *A Complete Scheme of School Music Related to Human Life* (Wellington, 1927) and the development of the country's first music syllabus in 1928.[9] Tayler energetically dictated all aspects of the syllabus at all levels. The gramophone figured prominently in his *Scheme* as a tool that could 'provide for the child an environment of fine music which will insensibly influence his taste towards what is beautiful'.[10] The government subsidised the purchase of gramophones and records for schools from 1924 until 1931, and Tayler's *Scheme* was intended to instruct teachers in the correct way to use the technologies and to teach musical appreciation.[11] Mechanical music, so often derided by high-culture zealots, was for Tayler an ideal gateway to an understanding of music, although he thought that only certain types of music were to be thus appreciated.

Tayler's *Scheme* spelled out the uses of the gramophone for each school grade in fine detail, with suggestions for an appropriate age- and level-specific repertoire. Special lessons could be developed that concentrated on musical pieces which had been heard many times through repeated playing. Tayler stressed that a record should not just be put aside if it made no immediate impression on the students: 'It should be heard many times, and an endeavour made to enter sympathetically into its purpose, period and nationality. Many records thus treated will be found to increase steadily in interest as they become more familiar.'[12] The logic behind this repetition was that familiarity with the forms and structures of classical music would lead to correct music appreciation – and to the correct habits and outward forms of disciplined and attentive listening.

Musical appreciation was also to be developed as part of other areas of the school curriculum. The appendices of Tayler's *Scheme* listed records (along with their catalogue numbers) that could be used in conjunction with studies of literature, history and geography.[13] Students studying the works of Rudyard Kipling might hear Peter Dawson singing 'Rolling Down to Rio' (HMV B1475) as they memorised the poem 'If'.[14] Budding historians of the French revolution could enjoy Albert Goosen's rendition of 'The Marseillaise' (Columbia 2555) while learning about regicide.[15] But teachers who contemplated illustrating a geography lesson about New Zealand with music were informed that:

> Like the other colonies, New Zealand has not yet developed a national music. Records of the older Maori (native) music are not available. It contains much repetition, small intervals, and a characteristic sudden drop in pitch at the end of sentences. The later Maori music, strongly influenced by European and missionary music, is graceful and happy, but not lacking in strong and characteristic rhythms.[16]

Tayler recommended recordings of 'Pokarekare Ana' and 'Hine e Hine' performed by Hohepa Porourangi, 'Waiata Maori' by the New Zealand singer Rosina Buckman, and 'God Defend New Zealand' as rendered by the Australian baritone Peter Dawson, complete with a speech by the New Zealand prime minister W.F. Massey about the British Empire. Thus geography, music and imperial pride were combined in one lesson.

The music outlined in the *Scheme* was overwhelmingly drawn from the canon of European composers. It ranged from Bach to Bizet, along with Elgar and Sibelius who were mentioned as two of the few modern composers worthy of being part of music education. Some folk songs and

TEACHING SOUNDS

A group of pupils in the healthy outdoors at a Methodist camp in Nelson – with their gramophone. There was probably very little jazz dancing going on here, at least while the teachers were watching. Nelson Provincial Museum Pupuri Taonga o Te Tai Ao, 311171

children's songs were also listed, but the recommended recordings tended to be performances by trained singers rather than unschooled folk musicians.

The emphasis on traditional art or high-brow music began from a child's first days at school. For the very young, Tayler's recommendations included nursery rhymes and 'anything which can be paralleled in the child's experience or imagination, fairies and giants, animals, toys, soldiers marching, people dancing etc. but avoid music of the cabaret type, instruments of ugly tone, comic songs and other trash'.[17] Tayler was adamant that 'music of the cabaret type' had no place in the classroom. The 1928 syllabus prescribed the correct use of the gramophone and the musical forms that were to be heard: '[It] must not be used for the repetition of the worthless music of the cabaret and variety show, but should build up a taste for music that is the expression of fine feelings, wholesome joy and fun, and sincerity of artistic expression.'[18]

By teaching children the discipline of attentive listening, albeit in three-minute chunks, they were to be safeguarded against the moral failings expressed by the sounds of pop music and the spontaneous and wild movements of jazz dancing.

The gramophone could also be used in more direct ways to control young bodies through dancing, exercise and drilling. Ted Harrall recalled the use of the gramophone as an exercise machine for both body and mind when he was a Christchurch schoolboy during the late 1930s: 'The gramophone was used for marching into assemblies and also for physical education. At Shirley Intermediate School there were no musical instruments. A Mr Beaumont took singing, and he played gramophone records of the classics.'[19] At Keripehi School on the Hauraki Plains the gramophone also played a role in exercise and discipline, as Margaret Coyne remembered: 'Drill was very important at school and it started off the day for everyone. After drill you had to march into school to the gramophone record every morning.'[20]

The drills and marching to the sound of a machine were part of a strategy to build better citizens able to withstand the debilitating effects of modern technologies by using those very technologies. New Zealand historian Caroline Daley has pointed out how physical exercise was seen as desirable by governments around the world during the 1930s, to strengthen muscles made flaccid by the depredations of cars, telephones, movies and radios. Increased leisure time did not necessarily make for better bodies, and schools were good places in which to begin inculcating healthy habits for the benefit of the people and the state.[21]

Along with strengthening children's souls through proper musical appreciation and hardening their bodies with rhythmic exercise, the gramophone was also used to guide their writing and speech habits. According to the educational philosophies that Tayler's *Scheme* was designed to uphold and reinforce, communication was to be correct in both sign and sound.

Myrtle Fraser-Jones perfected her penmanship, a vital part of 1930s education, to the sound of a 78:

> Mr Brown had the old gramophone. He would put the record on and we wrote to the music and that is how he taught us to write properly. He would have a certain letter that you would have to do on that line. It would be A and then another A with the emphasis on the down beat all the time and then the B's and so on. He used the beat of the music to make us write straight down with the first stroke.[22]

The gramophone also guided the speech of the nation's children, although it was sometimes seen as hindering correct usage rather than helping. A.N. Fitzgerald, a retired British schoolmaster, visited New Zealand in 1934 and commented in the *Education Gazette* on the purity or otherwise of the spoken word as he found it. Overall he was impressed with its quality.

However, there was some decay in the speech of the 'present generation' and he blamed it on 'misleading models' such as 'gramophone records of American speech'.[23] W.J. Mountjoy, a New Zealand teacher, voiced similar concerns about the sinister influence of modern technologies. He stressed that acquiring correct speech was more important than mastering the art of prose. Mass culture, in the form of talkies, the radio and the gramophone, was, he believed, the main culprit in the decline of the 'mastery of spoken English'.[24]

These men regarded new technologies as a threat to correct speech, but Tayler and other educationists also recognised that the gramophone could be a useful aid in the acquisition of desirable talking habits. Tayler's syllabus was mainly concerned with musical education and his references to speech concerned singing, but he called for 'Special attention to clearness of articulation, audibility, and purity of all vowels. Slovenly speech [is] as unpleasant and undesirable as bad workmanship in any other department of life. All ability to convey ideas intelligently depends upon right use of speech, or we are back at the level of animals.'[25] The gramophone was intended to assist in this. William Johnson was far more emphatic on this point than Tayler, considering as he did the whole range of education as mediated via the gramophone: 'Whatever shade of the King's English is spoken, it can at least be delivered clearly and with beauty, while it can become a living language when thought and spirit, and character colour the living words. Again the gramophone may be used to this end.'[26]

• • •

The important role of the gramophone in university music education is illustrated by the experience of the New Zealand composer Douglas Lilburn. Lilburn had first heard a gramophone at St George's School in Whanganui, which he attended in 1928 at the age of 13. He recalled: '[O]ne master owned a gramophone and introduced age-appropriate classics such as *Unfinished Symphony*, *Danse Macabre* and *The Carnival of the Animals*.'[27] Identified as a boy who displayed musical talent, he was encouraged to hear the approved musical canon.

Lilburn's listening continued at university. From 1931 the four university colleges had been given grants of US$5000 per year for up to five years by the US philanthropic organisation Carnegie Corporation as part of its policy of funding high culture. The colleges often used this money to purchase gramophones and records. Victoria University College acquired 'an excellent electric gramophone and loud-speaker, together with upwards of a thousand carefully selected records, covering a very catholic range'.[28] Canterbury College acquired an electric gramophone in 1935 and then another the following year, along with books on music and a thousand records. 'Gramophone recitals and musical appreciation lectures were popular features of the College's extra-curricular musical life'.[29]

Douglas Lilburn attended Canterbury College during these years and listened to as many recordings as he could. Such listening often involved score reading – a practice strongly recommended by educators and gramophone proponents such as Scholes, Tayler and Johnson as a way of increasing musical appreciation.[30] This was an important part of the education of higher-level music students like Lilburn, allowing them to develop a greater understanding of the structures and architecture of the music they heard. It was also a means of hearing modern music that was rarely performed in New Zealand. Lilburn described how he was still baffled and fascinated by Bartok's music even after listening repeatedly to recordings. He did not hear any of Bartok's work performed live until he went to London in 1937.[31]

The elite university cultural environment also fostered contacts with people who were interested in art music and had access to its sounds in less formal surroundings than those of the lecture hall or seminar room. Lilburn made friends with 'old Charlie Campbell in his St Alban's cottage' and listened to many works there through an acoustic machine. He heard Sibelius's *Symphony No. 2* and 'intuitively realised other directions related to my basic experience'.[32] Sibelius was to be a major influence on Lilburn's music. This type of informal but informative listening also went on in the flat of Lilburn's friend, the poet Denis Glover, often after an evening at the pub.[33]

Organised listening groups existed outside the formal education system, too, many of them designed to foster serious, disciplined listening and musical education, although some were highly informal in their approach. The Workers' Educational Association (WEA) and the British Music Society (BMS) Gramophone Groups were two organisations that promoted didactic listening, although for varying ends.

The WEA's music education was focused not on performance but on appreciation and understanding. It was not designed to subvert the established musical canon or modes of listening to it. Rather, it brought the ideas of Tayler, Johnson and other music educators who wished to inculcate middle-class and conservative ideas of music appreciation to working-class listeners.

Opera recordings complete with libretti and translations brought this exotic artform out of the opera house and into the home, no matter how remote. In New Zealand such recordings were an important part of the WEA educational system.

Northern Advocate, 7 September 1925, 14

The WEA operated outside the formal educational system, but it too used the gramophone as part of its teaching mission. During the 1920s James Shelley, director of tutorial classes, developed the Box Scheme as a way of supplying materials to people in remote areas who met regularly to study their WEA courses. A full course consisted of 24 boxes, each of which was passed from group to group, and contained lecture notes, prints, literary anthologies, play texts and gramophone records.[34] The scheme was highly successful, but as the Depression dragged on and the WEA's funds ran low, the material in the boxes began to suffer. A report in 1933 complained that many of the records were in need of replacement after years of hard usage.[35]

The Box Scheme was based on the traditional canon of classical composers (the first of its efforts included a section of notes and records on Beethoven and Schubert). However, this did not mean it was any less enjoyable than popular music. As one student later recalled:

> A great treat to most of us and a heavy load to carry, was our introduction to Grand Opera. About 10 or 12 large records (breakable in those days) and copies of the libretto were sent of 'Aida' and I have never forgotten that wonderful music. The gramophone had to be wound up for each side of the records. Our education and pleasure was hard work [but] for us with a thirst for knowledge and who lived in small inland towns, it was great.[36]

The middle-class listeners of the BMS Gramophone Groups also wished to extend their knowledge and appreciation of the approved canon. Founded in Britain in 1918, the BMS was established in New Zealand in 1932 by Valerie Corliss, a music teacher, on her return from the United Kingdom. Its goal was to encourage a taste for music in general, and it enrolled not only practising musicians and teachers but also anyone who felt music was important for national life. The New Zealand branch aimed to establish itself in centres throughout the country, and its activities were promoted through the periodical *Music in New Zealand* (*MNZ*), published from 1931 until 1937. The journal's editor Vernon Griffiths was, like Tayler, a British import. Born in 1894 he came to New Zealand in 1926 and took up a position as lecturer in music at Christchurch Teachers' Training College. In 1932 his appointment was retrenched owing to the Depression, and he became music master at Dunedin's King Edward Technical College, where he developed a widely copied programme of musical education. He became professor of music at Canterbury University College in 1942 and remained there until his retirement in 1962.[37]

An untiring zealot in the cause of British music, Griffiths believed that it had a large role to play in New Zealand's cultural life. He wanted to model New Zealand music education entirely on British lines. As Rachel Hawkey has noted, he was an Anglophile, a cultural conservative who was not impressed with modern music and its technologies.[38] *MNZ* was unashamedly highbrow, and the activities of the BMS Gramophone Groups followed this tone. Just as New Zealand's book clubs and film societies strove to maintain correct and serious cultural attitudes to the media they were concerned with, so the Gramophone Groups approached listening to elite music in a high-brow and didactic manner.[39]

The reports of the Hastings Gramophone Group give the flavour of these meetings, which typically consisted of a talk on a musical topic, accompanied by recordings. They seem to have been conducted very seriously. There was no Two-stepping or Fox-trotting at these social gatherings:

> The August meeting of the Gramophone Group was held in Mrs Christie's home in Havelock North, when the composer chosen for the evening's music was Beethoven. A most interesting paper was read on the composer's life and works, and amongst the gramophone records heard were the 'Emperor' Concerto played by Schnabel, the 'Archduke' Trio, and the string Quartet, Op.59, No.1. All the members present were most enthusiastic as to the possibilities of these gramophone meetings.[40]

The Gramophone Group gave an entertaining evening at Miss Dalton's residence on 1 September 1933. Mr Ewan Campbell had been asked to speak about modern music, and his little address proved most original and comprehensive, given as it was with illustrative records which were interesting and original and provoked animated discussion about modern composition.[41] The group revisited Beethoven for their final meeting of 1935 when they heard a lecture as well as piano sonata recordings performed by Walter Gieseking and Artur Schnabel.[42] The Hastings Gramophone Group was still active when *MNZ* ceased publication in 1937.

Groups in towns such as Hastings, Masterton and Nelson met in members' houses. Those in larger centres sometimes used other premises. The Wellington group held a meeting on 19 April 1934 in Begg & Sons' salon, where they heard a special HMV recording of Schubert's 'Winterreise' (Winter Journey) performed by the German baritone Gerhard Hüsch with Udo Müller at the piano.[43] The Dunedin group had a special treat in 1935 when they heard 'the gramophone presented to the Otago University by

> FREE.   FREE.
> GRAMOPHONE RECITAL.
> GRAMOPHONE RECITAL.
>
> TOWN HALL, AUCKLAND.
> MONDAY NEXT, APRIL 22,
> AT 8 P.M.
>
> A CHOICE PROGRAMME OF
> GRAMOPHONE RECORDINGS
> Will be Played on the
> BRUNSWICK ELECTRIC
> PANATROPE.
> The Following Artists Assisting:—
> MISS MOLLIE ATKINSON, Soprano (pupil of Madame Mary Towsey).
> MR. HARTLEY WARBURTON, Baritone.
> MISSES ISABEL BROOK, LAURIE HAMMILL, and JOYCE HOWIE, and MR. ERIC MIDDLETON, Dancers (pupils of Miss Cecil Hall).
> Accompanists:—MISSES TUI WARD and EDNA WARBURTON.
> Stage Manager   Mr. Karl Atkinson
>
> Tickets at all Howie's Shops. Box Plan at Howies, Ltd., 292, Queen St (opp. Smith and Caughey)
>
> No Booking Fee or Admission Fee.
> Presented by Howies, Ltd.

This 1929 free gramophone recital promised a night of varied entertainment. The Brunswick Panatrope was one of the exciting next generation of electric gramophones that added new dimensions to recorded sound. New Zealand Herald, 20 April 1929, 14

the Carnegie Corporation – a magnificent gift, with the addition of a large and valuable library of records'. The programme included works by Bach, Schubert and Brahms.[44]

Other types of semi-formal listening groups were made up of people attending gramophone recitals staged by retailers. These were held regularly in cities throughout the 1920s.[45] They were intended to advertise new machines and recordings, and could be classified as a form of didactic listening. Usually held in a theatre or town hall, the recitals normally consisted of live performances and recordings, preceded by an address from the stage concerning the recordings to be played and the machines they were played on. Speaking at a gramophone recital in Wellington in 1926, C.J. McKinnon, the *Dominion*'s record reviewer, noted some of the effects of the gramophone on musical life:

> The speaker concluded by emphasising the remarkable influence recorded music was having upon the musical culture of the general public. The most striking testimony to the power of really good music to assert its superiority over inferior music was the increasing preference of the public for the former. What used to be considered the exclusive preserve of a narrow circle of high-brows was now popular enjoyment for the multitude – good music had at last come into its own.[46]

By demonstrating the capabilities of gramophones and calling for disciplined listening, these recitals played an important role in normalising the technology. Admission was usually free, so they were also a cheap evening's entertainment.

Tone tests, developed by the Edison Company to market its Diamond Discs, were a variation of these recitals, and also found an enthusiastic audience in New Zealand.[47] They constituted a type of sonic Turing Test.[48] A singer or instrumentalist would alternate performances with a machine from behind a curtain, and the audience was invited to tell the difference between the player and the person.[49] Such recitals generally featured a large amount of classical music, and were heard as if they were live performances. Audience members dressed formally and behaved in the ways expected of serious listeners at a concert.

• • •

Along with these structured forms of public, didactic gramophone listening there were highly informal, pleasurable and semi-private groups that used the gramophone for educational purposes.

Arthur Pearse, New Zealand jazz broadcaster, was a serious-minded fan who listened to jazz as intently as others listened to classical music.[50] He sought out the company of like-minded souls and thought he had found such a group in 1932 when he heard about a gathering of dance musicians who met after hours at Jack McEwen's shop in Manners Street, Wellington, to hear the latest sounds on a portable gramophone. However, he was disappointed to find that other members of the group preferred to try to play along and learn from the records rather than simply listen to them. Their casual approach and the boisterous tone of the gatherings were not the sort of disciplined and serious listening Pearse desired, so he stopped attending.[51]

The musicians at these meetings were learning from the records in different ways. While Lilburn and others used gramophones to supplement the musical scores that were a central part of their education, those working in the popular (so-called 'low-brow') music field learned from recordings, radio, movies and the people around them. Tex Morton, New Zealand-born country singer, was 15 when he left home with a guitar and 'a few chords and tunes learned from a neighbourhood family of Māori; itinerant sailors and circus hands; and strumming along to American gramophone 78rpms of Jimmie Rodgers, Carson Robison and the Carter Family'.[52] Morton invented himself as a successful hillbilly singer during the 1930s on the basis of this musical education.[53] Another performer who learned the new music of hillbilly was Johnny Cooper, 'The Maori Cowboy'. He grew up during the 1930s near Wairoa and learned songs from records and films.[54] Macnamara's Family Band was active in Otago during the late 1930s: 'they were able to play a lot of the sheet music but a lot of music was learnt by ear'.[55]

Learning music from records which could be played again and again became the preferred way for most popular musicians to acquire songs, enabling them to learn by imitation and repetition. Published music scores often had little bearing on what was heard on the recordings, especially in jazz, blues and country, and vernacular music in general, and many of these musicians could not read music at a high level. Their need to do so lessened as the piano was superseded by instruments such as the guitar. Indeed, in jazz music, reading was often felt to place restrictions on improvisation.

Musicologist Mark Katz has pointed out that jazz reverses the dominance of the score by making the performance the 'primary text', with notation as a subsidiary interpretation.[56] This was true of much of the music that became popular in New Zealand during the 1930s, and these ways of learning formed a new approach to musical education made possible by audio technologies that opened up the performance of music to many people. The gramophone did more than teach people, however; it also moved them in all sorts of ways.

Chapter 5

MOVING SOUNDS

•••

The joys of jazz were not just for the players learning riffs from records. The new technology brought the pleasures of listening to the living room, where music could be enjoyed far from the crush of a crowded bar or the formality of the concert hall. Recorded sounds moved emotions and bodies. People used them to forge new bonds and strengthen old ones. A record might be used for consolation or comfort, or for cutting capers on a dance floor or a living-room rug.

Classical music fans benefited during the 1930s from the decision by many record companies, particularly European ones, to issue complete sets of works. Previously classical works had been cut up, rearranged and re-scored to suit the limitations of recording equipment such as short playing times and limited frequency response. With improved technology it became possible to make recordings that were faithful to the score.[1] The members of the BMS Gramophone Groups and their ilk devoured complete sets of works such as the Beethoven piano sonatas and the Mozart operas with high-minded appreciation and enjoyment.[2]

Individuals such as Lindsay Buick, author of *The Romance of the Gramophone*, also enjoyed this serious music.[3] But classical music was becoming a minority taste as gaps opened up between types of music. A British dealer lamented this in the late 1920s: 'Out of twenty-one complete works, only five have achieved satisfactory sales; of a total of nearly two hundred records of strictly classical music, only thirty have proved commercially successful'. By contrast, a theme song from a talkie had sold between 2000 and 3000 copies in a matter of weeks.[4] There are no comparable figures available for New Zealand, but it would seem likely that the same pattern prevailed.

•••

A thousand flowers of popular music bloomed between the wars. New forms such as jazz, hillbilly, blues, tango, polka, along with dance crazes and songs from the talkies, offered global audiences a feast of popular music that called forth myriad listening styles. The same piece of music might be listened to in a variety of ways depending on circumstances, company and mood. A chorus, for example, might capture attention while a verse was ignored in favour of conversation. Dance music made for even more complicated and unpredictable listening styles, depending on skill and situation. Dancing was very popular throughout the late 1920s and the 1930s, and the gramophone, along with bands made up of musicians made redundant from theatre orchestras by the talkies, played a large role in this pastime.[5] Record companies and gramophone manufacturers were quick to grasp the potential of recorded sound as a way for people to learn dances at home and practise any new steps that came their way.

Records provided dancers with the latest sounds and styles from around the world that a piano player could not supply. They set dancers in time with all the most fashionable steps and styles.[6] Many of these dances were invented by songwriters, music publishers and record companies to cash in on the global dance fashions.

The song 'Chalita' and the 'Tile Trot' dance from 1929 are typical examples. 'Chalita' was a tango written by film director and composer Victor Schertzinger in 1928. The song was highly derivative of the popular 'Yale Blues' of 1928. The 'Tile Trot' dance was itself a version of the Tango, and the Columbia label promoted it heavily as the next big thing. The popular newspaper *Truth* devoted two articles – complete with photographs – to teach the steps, and assured its readers that these were 'well within the scope of the average dancer'.[7] The promotional material used by Columbia stressed the novelty of the tune and the ease of the dance, and also promoted a public demonstration of both at Auckland's Strand Theatre. Both song and dance were also featured at the very popular Auckland night spot 'The Pirate Ship'.[8] The record stores' advertising of the song, which ran until the end of May 1929, emphasised the value of learning the dance in private. Watching professional dancers demonstrate the Tile Trot was one thing, but it was easier, more fun and less embarrassing to learn it at home. And it sold more copies of the song too.

This was another form of educational listening made possible by recorded sound. It may not have appealed to some of the Gramophone Society members who listened to records as if at a concert: attentively, quietly and

Above and next page: Records offered dancers the latest moves and sounds from around the world. New Zealanders moved their bodies to the rhythms of global modernity, and the record companies cashed in on their manufactured dance crazes. Auckland Star, 21 March 1929, 14; 28 May 1931, 10

Any Time is Dance Time

if you own these *Music by World-famous Orchestras.*

Columbia & Regal DANCE RECORDS

Columbia 4/- each

DO 358	On a Little Balcony in Spain (Foxtrot). When You Were My Sweetheart (Foxtrot). Jack Payne and his B.B.C. Orchestra.	DO 348	Wedding Bells are Ringing for Sally (Waltz). Debroy Somers Band. The King's Horses (Foxtrot). Ray Starita and his Band.
DO 347	Headin' for Better Times (Foxtrot). Just a Gigolo (Foxtrot). Ted Lewis and his Band.	07043	Body and Soul (Foxtrot). Something to Remember You By (Foxtrot). Paul Whiteman and his Orchestra.
DO 323	Down the Lane to Home Sweet Home (Waltz). Songs I Heard at Mother's Knee (Foxtrot). Ray Starita and his Band.	DO 317	I Miss a Little Miss (Foxtrot). Cheerful Little Earful (Foxtrot). Ben Selvin and his Orchestra.
DO 343	Yoi! Yoi! Mr. Cohen (One Step). What's the Matter, Abie? (Foxtrot). Jack Payne and his B.B.C. Orchestra.	DO 316	You're the One I Care for (Foxtrot). Tears (Waltz). Mickie Alpert and his Orchestra.

REGAL (Only 2/6)

G 20961	The Seven Step. The Peanut Vendor (Foxtrot). Denza Dance Band.	G 20994	There's Something Missing in Your Eyes (Foxtrot). Overnight (Foxtrot). Smith Ballew and his Orchestra.
G 21010	Little Sweetheart of the Prairie (Waltz). Don't Forget me in Your Dreams (Waltz). The Cavaliers.	G 20879	Strolling Thro' the Tulips (Foxtrot). The Swing Song (Waltz). Stellar Dance Band.
G 21005	Reaching for the Moon (Waltz). Golden Gate Orchestra. Fleur D'Amour (Waltz). The Society Serenaders.	G 20945	What Good Am I Without You? (Foxtrot). I'm Tickled Pink with a Blue-eyed Baby (Foxtrot). Lloyd Keating and his Music.

Columbia

Obtainable from ARTHUR EADY, LTD., 112, Queen Street; LEWIS EADY, LTD., 192, Queen Street, and at Hamilton; FARMERS' TRADING CO., LTD., Hobson Street; JOHN COURT, LTD., Queen Street; CHARLES BEGG & CO., LTD., Customs Street East; CADDELLS, LTD., Queen's Arcade, Queen Street; THE BRISTOL PIANO CO., 49, Karangahape Road, and at Hamilton.
ARTHUR H. NATHAN, LTD., Wholesale Agents, Customs Street East, Auckland.

Buy a record and learn the latest dance in the comfort and safety of home – no embarrassing public gaffes – with the aid of the 'free brochure' that spelled out the steps. Unlike a pianist, the gramophone never tired of playing the record over and over until the dance was mastered. Bay of Plenty Times, 14 May 1929, 4

studiously. But the living-room dancers who played 'Chalita' again and again as they worked out the Tile Trot according to the brochure supplied with the record were learning important skills that were also fun and fashionable. Advertisements and articles in the press about dancing and the current hit songs were ways of instructing people in how to use the gramophone as a normalised part of their lives. And they worked.

Mary Findlay was a teenager on an isolated Otago farm during the late 1920s when she and her friends prepared for the district's New Year's Eve ball with the help of a gramophone. Mary had never danced, but with the aid of the gramophone she quickly learned to waltz and 'by the end of the week I could do the foxtrot, Valeta, Maxina and even the Schottische'.[9] These may not have been the last word in fashion, but she and many others were able to expand their social opportunities cheaply and conveniently through the gramophone and dance records.

Along with moving the dancing body, popular music heard on gramophones also moved people's emotions. Broadcaster Arthur Pearse, who had been disappointed by overly flippant jazz fans, was deeply affected by Bob Crosby's 'Woman On My Weary Mind' when he was looking for a theme to open his 2YA jazz show in 1937: 'He sensed something different about it ... He played other records from the parcel but the Crosby disc stuck with him, and the next morning and through the day, he kept thinking about the track. He had found what he had been looking for for so long – a tune that was catchy, recognisable, yet elusive.'[10] In a similar way a song heard by a couple at the movies or on the radio might be purchased on record and become 'their' song, a signifier and stimulator of emotions that could be repeated anytime, anywhere.[11]

The portable gramophone could also be a roving instrument of annoyance for some. Elsie Billings lived in Pipiroa on the Hauraki Plains, and remembered her Uncle Bill being devoted to his gramophone and prepared to take it anywhere with him. Laughter was part of the listening experience but Uncle Bill expected his novelty records to be listened to attentively:

> My Uncle Bill was a character. He had this jolly gramophone and had these records. I know one of them was 'The Laughing Policeman'. He had a leather case which held so many records and he would go round to different places for the evening. He went to O'Neil's one night and Mrs O'Neil said that it was awful and the worry of her life trying to keep all her kids quiet. Everyone had to sit down and listen to the music.[12]

Jazz and popular music were more likely to be sources of irritation. Some people considered jazz both a pathological symptom and a threat to cultural and individual wellbeing. Some objected to it as unmusical noise. Others heard the sounds of jazz, particularly crooning, as unmanly. Yet others connected jazz to wider concerns about technology and modernity. For some, it was even the devil's music. Preachers, politicians, parents and writers raged at length against its supposed immorality and decadence. The music educator E. Douglas Tayler memorably described jazz as 'a musical revolt against the mechanical routine and uninteresting occupation by which the average young man or woman has to earn a living in shop or office. It is a big blowing off of steam.'[13]

The idea that a taste for jazz was a pathological condition caused by the pressures of the modern world was expressed by a number of writers. Karl Atkinson, a New Zealand music critic, summed up this attitude:

> Why does the waltz fail to satisfy? Simply because the nerves of the present generation are in such a state that they are bored by silence. A healthy, normal animal, whether human or not, is not bored by tranquillity, rest, silence. A man in a normal state can sit all day fishing or drifting along with a small breeze. When his nervous health begins to fail, he takes to more exciting pursuits, among which is jazz.[14]

Against the diseases of the modern age, Tayler prescribed a prophylactic course of musical education and appreciation. He was concerned to save the physical and cultural health of young people before they succumbed to the ravages of jazz and the Two-step. This was a medicinal as well as an educational use of the gramophone. It was beneficial to the nation's health in an age of increased noise – 'the diabolical symphony of our present mechanical age', as a New York commission on city noise put it in 1929.[15] The febrile sound of jazz was one motif that Tayler and others were determined to excise from the symphony of New Zealand's soundscape.

Taking up the theme was *The Mirror*, a women's magazine catering for middle-class readers. The music it favoured in its pages was middle to highbrow and, above all, respectable – which jazz was not.[16] Jazz was 'the cult of exaggerated syncopation … empty of emotion and sentiment', one of its writers declared in 1931.[17] *MNZ*, which was aimed at musically educated elites, went even further in its criticisms in another 1931 article. Jazz was noisy, unmusical, and marred by bad singers (who were 'particularly nauseating to musical people') singing senseless words in American accents, according to this writer. Jazz mangled the works of great composers; the players had no technique: 'we have it inflicted on us over the air, in restaurants, blaring from gramophone shops in the streets, and from our neighbour's gramophone'. The magazine hoped it was a passing fad.[18]

Such sentiments by the moral and the 'musical' did little to diminish the popularity of jazz. Just as the few objected to the talkies and radio while the many enjoyed them and wove the sounds into their lives, so jazz provided excitement, fun and glamour to both its devotees and its casual listeners. Jazz was the sound of modernity and being modern was all the rage.[19] Myrtle Fraser-Jones recalled in 2004 Saturday nights at the Ngatea Hall: 'They would have Epi Shalfoon and his band come down and play. That was really music. You could sit and listen or dance with that music all night compared with what you get today.'[20]

Crooning, a vocal technique made possible by developments in microphone and amplification technology, was another aspect of jazz to arouse suspicion. Singers like the American Rudy Vallee used the microphone as an instrument in its own right, rather than as a vocal prosthesis.[21] But the objections to crooning were framed in terms of gender and musical ability. It was described as unmanly as well as unmusical.[22] Such fears reflected uncertainties around gender roles and performances in a time of great change.[23] The jangling sounds of modernity blaring from the new audio technologies challenged the traditional sounds of His Master's Voice.

Some of the gramophone's critics during the 1930s heard technological modernity as threatening to both music and culture. Such criticisms had a long pedigree: John Philip Sousa had raised fears about the gramophone as 'canned music' in 1906.[24] The composer Constant Lambert and the philosopher Theodor Adorno also warned against the dangers of 'canned' or mechanical music.[25] New Zealand journalists and educators expressed similar ideas, but there were differences in intention among these commentators.[26]

The New Zealand critics who decried the new audio technologies, and particularly the gramophone, did so because it was inimical to amateur music-making. Technology cut down employment for music teachers and musicians, and 'it is an open question whether we are recompensed by the vast volumes of sound'.[27] Leisure itself was seen as being mechanised and mass produced by records, radios and movies, and this affected the nation's creativity and artistic abilities: 'Mass mechanical amusement has taken the place of personal endeavour to the impoverishment of the individual.'[28] Listening itself was becoming a mechanical process. The gramophone brought art music to the masses, but also made 'great' works overfamiliar, meaning they were not heard with due reverence and seriousness.[29] Lambert lamented that there was no escape from mechanically reproduced music in public and that even if the music was high art, it was heard in unsuitable surroundings 'whether in town or country, in a motor car, train or restaurant, perched on a hilltop, or immersed in the river'.[30] High musical culture was de-sacralised by the ease with which it could be heard through the audio technologies of modernity. The sort of listening these critics defended against the ubiquity of mechanised, mass culture was disciplined and reverential. Popular music was inferior not only on musical grounds; it was unworthy of correct musical listening and performing habits.

Even gramophone enthusiasts such as Buick considered the machine a high-cultural device that reinforced established and elite ideas, both of what was heard and how it was listened to. For Buick, the social utility of the gramophone lay in its ability to bring 'good' music to all people in all parts of New Zealand. Hearing this music could strengthen home life against the threats posed by 'pictures, jazz parties and cabarets'. The gramophone would 'do much to inspire a love of music and banish the loneliness of the back-blocks'.[31]

The intent of Adorno's critique of mass culture and mass media was entirely different. Adorno's ideal listening may have been art music, and particularly avant-garde art music, but he was concerned with listening as a means of radicalising and liberating subjectivity from conservative cultural norms. Not even the WEA, seen by some as a force for subversion, went this far. It was probably not what the cultural gatekeepers of New Zealand had in mind when they criticised the ways in which the gramophone and other audio technologies were listened to in the age of mechanical reproduction.

• • •

Playing recordings rather than making them was the norm between the wars in New Zealand. The mainstream, commercial life of recorded sound, whether on record, radio or film, was based almost entirely around overseas sounds which promised and delivered a wide range of musical pleasures from opera or dance music at home to jazz on the beach – and brought with them their varied associations of glamour, sex, money and cultural capital.[32] New Zealand had no commercial record-making technology or infrastructure until 1949. The recordings that were made were usually produced in radio stations or at the larger music stores, and tended to be one-off acetate discs used in broadcasting or as private mementos. Few traces of these activities remain, as the discs were fragile and survived only 20 to 30 plays before becoming unusable.

Tex Morton made recordings in Wellington during 1932 that are generally acclaimed as the first recordings of country music outside America. These special one-off aluminium discs were never released commercially, although they did receive some radio airplay.[33] But local recording activity was sparse. As we will see, apart from recordings of Ana Hato made in 1927 and a series of recordings made in Rotorua in 1930, all of which used equipment brought over from Sydney, recordings of New Zealand performers were made in Sydney or London. Apirana Ngata's recordings of traditional waiata, made during the 1930s, were not heard publicly.[34]

The few commercial recording activities that related directly to New Zealand's indigenous sonic life during the 1930s were made by international recording companies and concentrated on commercial Māori music. These were part of a worldwide strategy that exploited local and regional musical talent to enhance sales in those regions and locales. Media historian Pekka Gronow has pointed out that the recordings made by the large record companies of Oriental music between 1910 and 1930 were of 'anything that might interest the primarily urban record-buying public'. Sales were not large in real terms but were 'quite enough to keep the business going, enough to establish a permanent relationship between performing artists and the recording industry, enough to create a regular record-buying audience, however small it may have been in relation to the total population'.[35]

The recordings made of New Zealand artists between the wars arose out of the competition between international record companies. Ana Hato was one of a group of Māori singers who were recorded by the Parlophone Company in a meeting house at Rotorua in 1927.[36] She was a respected singer and guide at Rotorua, and one of the performers who sang before the Duke and Duchess of York during their visit to New Zealand that year.[37] The master discs of the group were sent to a pressing plant in Sydney for manufacturing, and when the recordings became available in New Zealand they were well received, with reviewers particularly noting Hato's voice. One reviewer described the recordings as 'the greatest gramophone event so far as New Zealand is concerned. Every one of the eight records is worth having. In fact I am sure that few New Zealander gramophone lovers will be able to resist them.'[38] Hato and her cousin, baritone Deane Waretini, made further recordings for Parlophone in Sydney during 1929 and these also seem to have been popular, although no exact sales figures are available.[39] The Waretini dynasty endured. Waretini's son, also named Deane, was to have New Zealand's first ever number 1 hit sung in te reo Māori: the 1981 song 'Te Piriti' (The Bridge).[40]

The decision by an international company to record Ana Hato was based on the logic of commercialism. On one level, her initial recordings

Ana Hato and Deane Waretini were marketed by Parlophone as traditional Māori performers. Alexander Turnbull Library, Wellington, Ref: Eph-A-PHONO-1930-01-5

were souvenirs of the visit by a popular royal couple. On another level, Parlophone passed itself off as performing a cultural service with a policy of 'recording singers in all countries and of all nationalities likely to be of interest to its many patrons, irrespective of the cost entailed thereby'. While Hato's voice was untrained, it was so beautiful that it deserved to be heard 'in the principal opera houses of the world'.[41]

Parlophone's 'policy of recording singers in all countries' was not unique, however. Commercial recordings had been made in many parts of the world since the late 1890s as record companies realised that catering to local tastes and consumer desires was far more profitable than expecting one type of sound to suit all ears.[42] They shaped their goods to suit local tastes in the interest of profit, rather than through any desire to record and preserve the diverse music of the world.

Parlophone furthered its stake in Māori music by recording the Tahiwis of Otaki in Sydney during 1930.[43] The siblings Weno, Hinehou and Henare Tahiwi (Ngāti Raukawa) were part of a prominent and musical local family.[44] The record company publicised their recordings by positioning the group as folk singers whose performances captured the 'real essence, the spirit and traditions of the tunes of Maoriland'.[45] Along with Māori songs, the group also recorded popular music of the day, including four songs from musical films. The *Evening Post* review of the recordings commended the group's performances of 'Happy Days Are Here Again' (from the 1929 film *Paris*) and 'Somebody Mighty Like You' (from the 1930 film *Chasing Rainbows*).[46] Many of the songs were jazzed up, in keeping with the sounds of contemporary pop music.

The record company Columbia responded to the success of the Hato and Tahiwi recordings by sending the British-born Sydney-based record producer Gil Dech and a team of technicians to Rotorua for three months in 1930 to record Māori singers. Over 20 tiring days, Dech and his team made many recordings of the 27-strong Rotorua Maori Choir in the Tunohopu meeting house beside Lake Rotorua.[47] The resulting discs were manufactured in Australia and released on both sides of the Tasman. Described by music historian Peter Downes as 'a prototype for future Māori recordings', they were favourably received and seem to have sold well.[48]

The commercial recordings of Māori singers and the circumstances of their production illuminate the ways in which the imagined community of New Zealand in the 1920s monitored, maintained and controlled the noises that may have otherwise undermined or complicated the dominant Pākehā

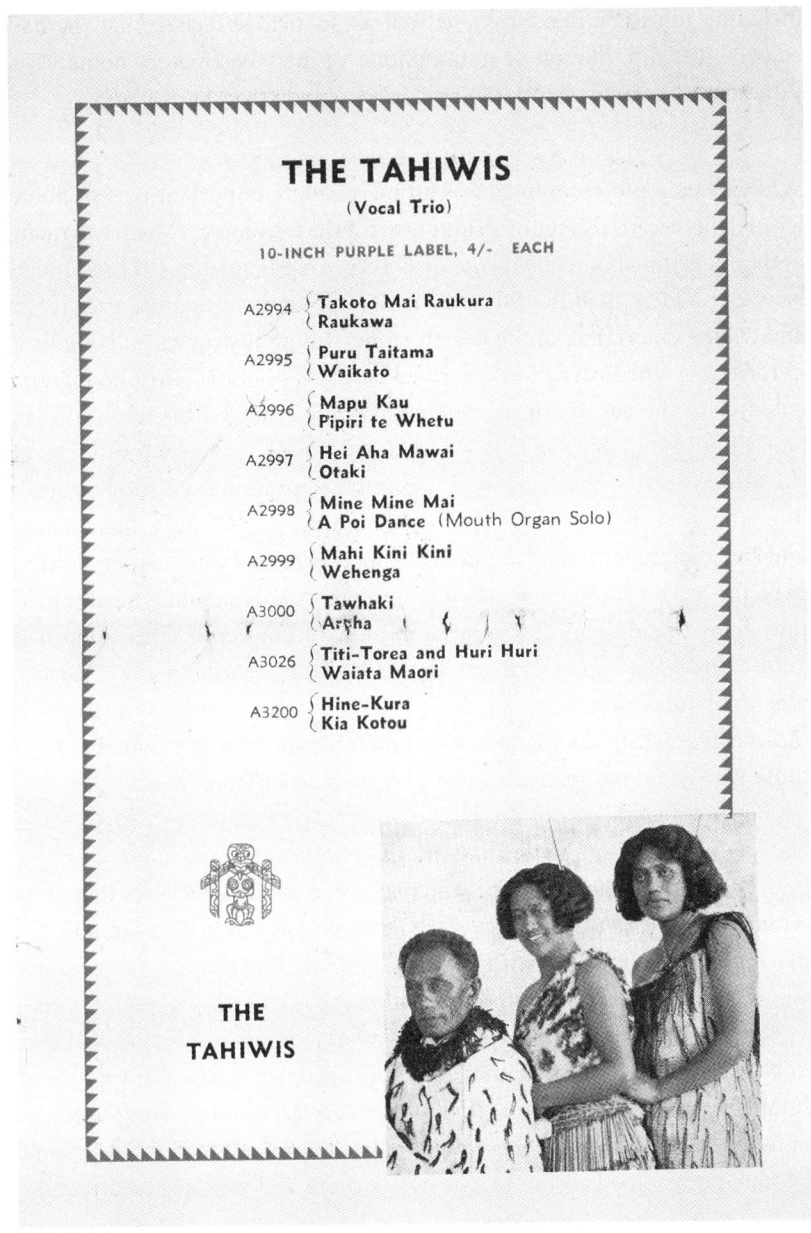

The Tahiwis were a respected performing family from Otaki whom Columbia marketed as 'relics of Maoriland'. Alexander Turnbull Library, Wellington, Ref: Eph-A-PHONO-1930-01-6

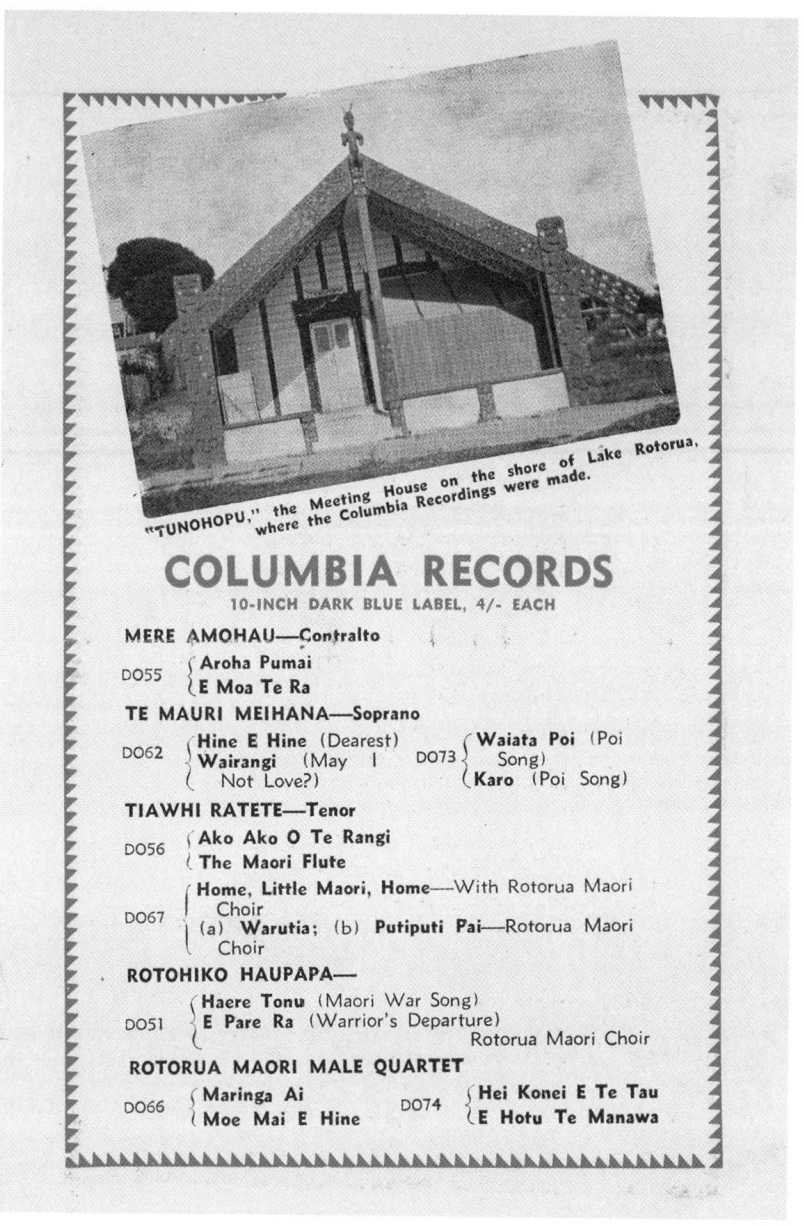

Columbia recorded many Māori singers at Rotorua over a three-month period during 1930. The records were marketed as authentic Māori music, with the famous meeting house Tunohopu, where the recordings were made, used to associate the music with ancient Māori traditions. Alexander Turnbull Library, Wellington, Ref: Eph-A-PHONO-1930-01-3

Ernest McKinlay was a Pākehā singer who recorded many Māori songs. Dressing him in traditional Māori garb – as for Hato and the Tahiwis – may not have been appropriate, so Columbia associated his music with traditional Māori culture through stock images. Evening Post, 10 November 1928, 24

conception of nationhood, and particularly the roles allocated to Māori within this 'national imaginary'.[49] According to Claudia Lemke:

> Māori did not pose a threat in the role of cheerful entertainer. The delicate role of consensus could be strengthened by music and laughter. In both roles, Māori could be aestheticized and commodified, more often than not with an eye for profit. Pākehā took pride in Māori achievements and what they deemed good 'race-relations', so long as Māori did not pose a threat to the nation-myth of being one people.[50]

The contrast between the sounds of commercially recorded Māori performers and those of traditional Māori music echoed this point. The commercial recordings were framed in the sonic styles of pop music and the repertoire consisted of pop songs and very well-known Māori songs. These followed European harmonic, rhythmic and melodic structures, conventions and patterns. They were assimilable for the ears of the Pākehā majority in ways that traditional Māori music was not. They reinforced the idea that New Zealanders were one harmonious people singing from a single score.

The visual marketing of the Māori records presented the performers in traditional attire and settings. The association of the performers with the trappings of pre-European Māori culture was designed to add to the authenticity of the music. In the case of Ernest McKinlay, a Pākehā performer of Māori songs, a portrait was dispensed with and stereotypical images of Māori life were used. These included a meeting house, carvings and a geyser, similar to the cultural motifs presented to tourists at Rotorua in the form of model villages and guides (such as Ana Hato). The sounds of the records were a safe and sanitised version of the sounds of contemporary Māori life. They constituted a fantasy form of cultural tourism.

The hybrid of Māori and pop music recorded by Hato and the Tahiwis may have been marketed and reviewed as 'authentic' Māori music, but was not universally admired. Politician and anthropologist Apirana Ngata had been involved in recording traditional Māori music from 1919 and continued this work throughout the 1930s. He openly questioned the authenticity of the commercial recordings:

> They were not true Maori songs. They were just as much Maori as the Hawaiian songs were Hawaiian. Merely English and other songs translated into Māori or Hawaiian. On the gramophone records they could not find one single song that might be called a Maāori song, but only songs

translated into Maori. There was a waltz or a jazz time, and a song sung in Māori. These were the songs on the records. But they had to go further back to get the true songs, incantations and chants of the Maori.[51]

Along with the social, cultural and economic divides that separated Māori from Pākehā in the 1930s, sound functioned as another boundary marker. For some, such as Ngata, it marked a line between 'authentic' and fake Māori culture. For cultural gatekeepers such as journalists and record reviewers, the sounds of the recorded Māori groups were proof that New Zealand's culture was homogenous.

Crossing such sonic boundaries could be hazardous, as Ernest McKinlay found out. McKinlay had been part of the 'Kiwis' concert party attached to the New Zealand Division during World War One and also worked as a professional singer in Great Britain.[52] He recorded a series of Māori songs for Columbia from 1928 onwards as part of that company's response to Parlophone's success in this area. These were well reviewed and, judging by the number of extant copies of his records, seem to have been very popular.[53] He related in his autobiography that he had learned some Māori songs without knowing what the words meant. He performed one of these in front of the renowned guide Maggie Papakura and was surprised when she made no comment or applause but merely raised her eyebrows. Later he learned from a friend that 'if you sing this song in English, you get put in jail'.[54]

• • •

The recordings made by Hato, the Tahiwis and the Rotorua Maori Choir were the first stirrings of a local industry. They were local performers making local music, some of which was recorded locally although much of it was laid down in Australian studios. But it would be more accurate to hear them as extensions of the global music industry that was in place by the 1930s. Since the early 1900s record companies had recorded local sounds to sell back to local audiences. The Gramophone Company producer Fred Gaisberg made extensive journeys through Russia, India, China, Japan, and Myanmar from 1901–03 with the express intention 'to open up new markets, establish agencies, and acquire a catalogue of native records'.[55] This was not philanthropy or ethnomusicology, it was a quite ruthless commercial strategy designed to maximise market reach and corporate profits. This applies equally to the Māori recordings.

For New Zealand audiences, though, these records were, and are, sonic taonga. They were very popular in their day and have been reissued many times over the years on LP, cassette and CD. For local listeners these records are not just mass-produced factory objects but important and precious art works that help us understand what it is to be in New Zealand. The music itself symbolises the complex cultural connections and exchanges that make New Zealand what it is now, and stands as a document of a particular moment in the country's cultural history.

By the end of the 1930s New Zealanders were avid consumers of the records and machines churned out by the global record companies. From the 'wonderful noises, not at all Shakespearean' first heard during the 1879 phonograph exhibitions, the recorded sounds New Zealanders listened to had become a cacophony of voices and music of all sorts. Local listeners were hearing the world through their gramophones, and the sounds of New Zealand were taking their place as a small but important chorus in this global din of recorded sound. Listening in was as important as listening out.

Part II | RADIOS

Chapter 6

RECEIVING RADIO

•••

Eric Battershill was just 16 when he appeared before the Hastings District Court in 1913. He was facing serious charges and could have been fined £500 if found guilty. His crime: 'he did establish a plant for the purpose of receiving communications by wireless telegraphy without having obtained the consent of the Postmaster-General'.[1] Battershill had built from scratch a radio receiver that could receive messages from 30 miles away but could not transmit. What he had not done was obtain a licence. Under section 4 of the Wireless Telegraphy Act 1903, anyone who constructed equipment capable of sending or receiving messages via wireless telegraphy 'without having first obtained the consent of the Governor in General is liable to a penalty not exceeding five hundred pounds' and their equipment would be confiscated.[2]

H.A. Cornford, prosecutor for the Crown at Battershill's trial, pointed out the seriousness of the case: 'What the boy was doing for his own instruction might be done by others for reasons serious to the country and Empire such as the interception of wireless messages from overseas. Such plots might produce results of the utmost gravity.'[3] The defence scoffed at this idea and maintained that the wireless regulation legislation served only to stultify the talent of a scientifically minded youth who was interested in all things mechanical and modern. No harm had been done and Battershill had been encouraged in his technical endeavours by his teachers. The presiding magistrate, however, pointed out that the Act was designed to prevent messages falling into the wrong hands. Someone with the sort of equipment that a schoolboy like Battershill could build might intercept private messages as well as those relating to matters of government. The law had to be obeyed. In the event, he discharged Battershill on condition he did not use the apparatus again.

The Eric Battershill case was widely reported throughout New Zealand.[4] There was a national interest in the exciting modern technology of radio,

Eric Battershill with an Oldsmobile 6 cylinder Mitchell in 1914. Eric's knowledge of wireless (and cars) came in handy during the Great War when, among other roles, he was driver and a radio operator for the Australian army. Knowledge Bank Hawke's Bay Digital Archive, 876/1297/37309

although 'radio' in 1913 was not radio in the modern sense of the term. As a one-to-many broadcasting system that transmitted audio in the form of speech and music, radio got underway during the early 1920s after a spate of technological innovations, some of which had been developed during World War One.[5] Prior to this, 'wireless telegraphy' was the phrase used to denote radio transmission. Messages were sent in Morse code from one point to another (one-to-one) in the same way telegraphs were transmitted from a sender to a receiver. In theory, each receiver would also be a transmitter, although amateurs such as Battershill could often only receive messages. The model of wireless telegraphy as point-to-point transmission of Morse code became the province of the Ham radio hobbyists ('DXers') as broadcast radio entered New Zealand's homes during the 1920s.

Radio transmission in Morse code was heard through earphones. The sparse and sporadic 'dits' and 'dahs' of tapped-out Morse messages occurred against a constant background of the whooshing, buzzing, howling, humming, clicking, sometimes screaming sounds of electromagnetic waves interacting in Earth's atmosphere.[6] It required patience to find any kind of coherent radio signal with the sort of equipment possessed by amateurs. It also required experience and knowledge to identify human-made transmissions and then decode them.

Calling this audio experience 'listening to radio' deafens us to the cultural meanings and significances the medium once had. The histories of all media involve complex ideologies, practices and technologies that do not have clear-cut boundaries. In the evocative phrase of media scholar Carolyn Marvin, 'Media are not fixed objects: they have no natural edges.'[7] This idea is highlighted by the varied ideas about radio held by New Zealanders before the early 1920s. For some, radio was a threat to the nation's security, while for others it was an interesting set of technical problems. Some thought of radio as a path to a future utopia and others regarded it as an ethereal link with the afterlife. As an exciting new technology of an exciting new century, radio was in the air in metaphorical as well as literal ways. Very few New Zealanders had heard a wireless telegraph set before the broadcast era began.[8] But many knew about the technology through newspapers, magazines, books and political debates. People heard *about* wireless telegraphy before they heard it. And this latest technology of modernity meant many things to them before it became radio in the sense we understand the term today.

• • •

The signals left by much of the wireless activity in New Zealand before the early 1920s are faint, and often ambiguous. Amateur enthusiasts working outside the law, such as Battershill, did not publicise their activities. Most wireless users were young men and boys who cared more about building wireless sets and making them work than documenting their activities for posterity. If wireless telegraphy was constantly in the news, the amateurs were usually noticed only when they were in court, although as we will see, some were noted for their impressive achievements. By contrast, the official, governmental use of wireless has left a much clearer record, and various scholars have written about the institutional history of the New Zealand wireless telegraph service in peace and war.[9]

Regulation was New Zealand's official response to the nascent communication medium. The government already controlled telegraph lines and telephones, and wireless was regarded as a similar communication technology.[10] The single page of the 1903 Wireless Act allowed for the government to set up wireless telegraphy stations, and threatened any unauthorised station operators with confiscation of their equipment and a fine of up to £500.[11] The main point of these harsh penalties was to deter competition. The government monopoly over communications was explicitly stated several times during the passage of the Act.[12] Objections that the proposed Act might stifle scientific research and experimentation were dashed when Sir Joseph Ward, Postmaster-General and Electric Telegraph Commissioner, pointed out the threat to national security posed by the nature of wireless communication. Anyone who set up a receiving station might be able to intercept messages and this 'would destroy the value of the whole system'.[13] The messages that maintained the smooth functioning of national and imperial government would be protected from unauthorised reception through the control of postal services and telegraph lines. The spectre that haunted the 1903 Act was the fear of some hostile foe listening in to state secrets. Keeping a potentially lucrative communication system as a government monopoly was just as important, but invoking threats to national security was an unbeatable political argument.

The framers of the 1903 Wireless Telegraphy Act envisioned the new media as another version of the more familiar technologies of the letter and the telegraph. It would be used to send messages from one person to another. For the New Zealand government, those listening to wireless telegraphy were not a collective mass audience but rather the individuals the messages were aimed at – ideally government employees who would be managing the wireless and wired telegraph systems in the same manner. This control was made explicit by the Post and Telegraph Act 1908, which restated the contents of the 1903 Act but brought wireless telegraphy under the direct control of the minister in charge of the Postal and Telegraph Department.[14] Wireless telegraphy was a serious matter and best handled by the government, not amateurs experimenting at home.

Official discouragement of amateur wireless operations continued in 1912 with an order to dismantle all unofficial aerials. The government had started to construct a network of wireless stations and feared that the amateur signals might interfere with official transmissions.[15] The point was made even clearer by the Post and Telegraph Amendment Act 1913, which made it easier for the government to obtain a conviction for unauthorised wireless telegraphy. Individuals could now be prosecuted simply for building equipment without government consent if that equipment had the capability to receive or transmit signals. Whether it was actually used for these purposes was irrelevant.[16] But the Act did allow permits to be issued to suitable applicants who wanted to experiment with wireless. Canterbury College applied for such a permit, but the outbreak of war in 1914 stymied the approval process.[17]

The war years saw a complete ban on any amateur wireless activities. Private ownership of any wireless equipment was illegal, and at least one individual was imprisoned early in the war for operating such a device.[18] This strict regulation of activities by any person or organisation other than the government continued until 1920 when the bonds began to be loosened as wireless telegraphy gradually started to be heard as radio broadcasting.

The tight regulation of unofficial wireless activities enforced by the government is the background noise to the reception of wireless in New Zealand from 1900 until the early 1920s. But the treatment of unlicensed wireless operators was inconsistent. On the one hand, such operators risked criminal conviction, large fines and even imprisonment. On the other, offenders were often let off with a warning, as in the case of Battershill. The government's monopoly divided wireless users into two groups: the clandestine experimenters, and the official users who worked for the Post and Telegraph Department at the wireless stations built by the government during the 1910s. Some members of the former group became members of the latter as wireless became more widespread.

The 1903 Act had signalled the New Zealand government's awareness of the new technology of wireless. In 1911, after some years of debate and information gathering, the first official radio stations began to be established.[19] Two powerful stations were situated at opposite ends of the country, one at Awanui, near Kaitaia in the North Island, and another at Awarua, near Bluff in the South Island. There were less powerful stations at Auckland, Wellington and the Chatham Islands.[20] Together they covered New Zealand's landmass and territorial waters.

For many people these installations provided the first physical experience of wireless technology, and were visible proof of New Zealand's modernity. The Wellington station's aerial ran for 90 metres between two masts, each 45 metres high. It was positioned on Te Ahumairangi in the Tinakori Hills overlooking the city and harbour. The Auckland station was located at the Chief Post Office in the middle of the city and used 12-metre steel masts to support the aerial.[21] A postcard from 1913 shows the second aerial mast being raised on the Chatham Islands in September of that year, and suggests the scale of the equipment required for long-distance wireless transmission. The aerial masts were 45 metres high, the same as at the Wellington station.[22] The two staff at the Chatham station were based in the house seen in the postcard and alternated work shifts. Every day the station was operated from 9am until 1pm, 3pm until 5pm and 7pm until midnight.[23] The other stations

Raising the aerial mast on the Chatham Islands, 1913. This engineering feat was another link in the wireless chain that guarded the security and safety of New Zealand's shores and shipping. Postcard, Alexander Turnbull Library, Wellington, Ref: Eph-B-POSTCARD-Vol-3-054-ctr

around New Zealand operated all the time. The men at these stations – and they were all men – sent and received messages that were mainly concerned with shipping. Their hours were spent listening attentively for any important signals. They were working with state-of-the-art technology in a new and constantly changing medium.

The technological glamour of wireless also stimulated amateurs such as Eric Battershill to experiment with their homemade equipment. Battershill was typical of the few unofficial wireless users active between 1900 and the early 1920s. The majority were male and often still at school. Most were interested in science and technology, and their interests were encouraged by their teachers and parents. Brenda Bell, the only woman known to be involved in early wireless activities, learned about wireless from her father Alfred Bell. She eventually became New Zealand's first woman amateur (or Ham) radio operator. Her brother Frank, also taught by their father, made history in 1924 as part of the first two-way radio conversation between New Zealand and Britain.[24] Radio had spanned the globe

Local wireless tinkerers may also have been encouraged by the example of the American Walter J. Willenborg. A student from Hoboken, New Jersey, Willenborg became a national celebrity when the *New York Times Magazine* profiled him and his wireless equipment. He became a symbol and a model for the young of America, a new type of 'boy hero' who, rather than subduing the dangers of the frontier, was able to use science and ingenuity to control technology. He used knowledge rather than physical prowess, and offered an alternative to the emphasis on bodily strength emphasised by the cult of fitness of the 1890s and 1900s.[25] New Zealanders also enjoyed the more abstract feats performed by Willenborg, whose exploits were favourably reported and commented on in the local press.[26] Indeed, interest in science and technology was 'curiously acute in New Zealand'.[27] The boy heroes of local wireless telegraphy were similar to their overseas peers in that their explorations of the ether were also explorations of a form of masculinity that valued technical knowledge and scientific skills over those of the sports or battlefield. This technocratic, white-collar masculinity questions the idea of a 'Better British' colonial masculinity built solely around physical hardiness and outdoor pursuits.[28]

Typical of the individuals involved in New Zealand's early wireless scene were the so-called 'three clever boys' of Dunedin. Sixteen-year-old Rawson Stark and 17-year-old Stanton Hicks and Cyril Brandon established three wireless stations around the city in 1908. None had a licence to send or receive wireless transmissions, but their achievement was favourably reported throughout New Zealand, and their demonstrations (during which Post Office telegraph operators performed the Morse coding) were attended by local mayors and officials.[29] On one occasion the boys received signals from a ship hundreds of kilometres away in Foveaux Strait. They even received a wired telegraph of congratulations from Prime Minister Joseph Ward.[30] He described the transmissions as bringing the 'highest credit upon the boys. I congratulate them upon the possession of such inventive faculties'.[31] These young men had openly flouted the regulations against unlicensed wireless transmissions but received accolades instead of prosecutions. The public nature of the event was one of the factors that kept them out of trouble.

They were also helped by the fact that Stanton Hicks' father was a journalist at the *Otago Daily Times*, and this ensured good publicity about the transmissions. Cyril Brandon worked for an electrical engineering company and Stark's father was one of Dunedin City's electrical engineers.

The wireless made by the 'three clever Dunedin boys'. The coil in the centre of the picture was wound by hand. They also built 13-metre tall aerials at each of the stations they established, with a cable leading to the transmitters. This was not a piece of equipment that might grace the parlour. Sir George Grey Special Collection, Auckland City Libraries, AWNS-19080924-11-2

These were clearly respectable young men who posed no threat to the city or the country. The press coverage emphasised the years of experimentation, tinkering and hard work that had gone into the construction of the wireless equipment. The hours they spent studying in the Dunedin Athenaeum were also an important part of this narrative. All the equipment was built from scratch using items that had been scrounged or donated, thanks in part to Stark's and Brandon's connections to electrical engineering in Dunedin.[32]

Eric Battershill and the 'three clever boys' were not the only New Zealand amateur wireless enthusiasts experimenting before the war. Many of them escaped official notice or were discreetly ignored by local postmasters.[33] But there was a sharp crackdown on amateur wireless activities between 1910 and 1914 as the government developed its own stations and wireless was increasingly used by shipping.[34]

In 1910, for example, the 'three bad boys' of Timaru – Alfred Hathaway, a P&T telegraphist, Burall Courtis and H.R. Mayo – had their wireless gear confiscated after their signals affected transmission between the warship HMS *Pioneer* in Timaru Harbour and HMS *Challenger* at Wellington.[35] In their defence, Hathaway pointed out that the *Pioneer*'s wireless operators had seen the trio's equipment and could simply tune away from its signal. He also mentioned the very short range of the gear.[36] There was support for the trio in the press, but to no avail.[37] Their subsequent application for a licence was turned down.[38] The government was taking wireless very seriously as the technology became part of naval and commercial shipping, and in 1912 all amateurs were ordered to take down their aerials. Some amateurs carried on their activities in secret, using various ingenious methods of hiding their aerials and apparatus.[39] With the outbreak of war in 1914, however, listening to wireless became a purely official activity. Only approved personnel in the armed forces, Postal and Telegraph Department and merchant shipping were allowed to send and receive wireless messages. The radio spectrum was tightly controlled and monitored by the government to prevent any unauthorised messages crossing from either side.

Wireless signals were heard in the internal sonic space generated by earphones. (Loudspeakers arrived with the advent of broadcast radio.) Postal wireless operators used earphones that were produced to a standard design. Amateur wireless operators had to make their own earphones or 'obtain' them from official sources. One early radio user recalled: 'The phones were of p. and T. [Postal and Telegraph] pattern, obtained from never mind where, suffice to say that they were obtained.'[40] The slightly swaggering tone hints at the sense of excitement that was part of the experience of early wireless. The New Zealand operators were flouting the law simply by building their equipment; they were rebels without a licence. And having acquired the earphones by any means necessary, listening through them was another thrill beyond most New Zealanders' ordinary experience.

Discussions about the use of earphones as listening devices have tended to concentrate on the ways in which they create private aesthetic spaces literally within the listener's head and bring a sense of privacy to individuals in public spaces.[41] Yet earphone listening began in medical and communication contexts. The stethoscope, the telephone and the wired telegraph made listening private and personal only in the sense that the sounds being heard were not propagated through public spaces.[42] What the experience of earphone listening generated for early wireless auditors was a form of listening that was attentive and discerning. The signals a listener sought to hear were often hard to separate out from atmospheric interference, let alone other signals. Earphones isolated the wireless listener from external sounds that might drown out the signals coming through the aerial, and highlighted sounds in a way analogous to the way a lens can sharpen perceptions of very small or very distant objects.

One account from 1922 gives some idea of what was heard through these earphones. When Ken Collins, chief technician at 2YA until 1949, recalled his first experience of radio before World War One, he described hearing through a 'faint high-pitched musical note in sharp staccato Morse code – "dit dit dit *dah*, dit dit dit dit". Suddenly the singing note died and immediately, much louder, a dry tearing sound took up the Morse code.'[43] These were the sounds of a Sydney radio station communicating with a ship in the Tasman. The characteristic 'dry, tearing sound' was made by the spark gap transmission system that early wireless used to generate signals.[44] A journalist hearing wireless through earphones wrote about 'scratching and spluttering' sounds. From this came a 'high, thrilling note' as a signal arrived.[45] But even if listeners could work out which of the varied noises was a Morse communication, the message was meaningless unless they could decode it, and this took training.

Clive Drummond was one of these listeners. Drummond, who became a well-known radio announcer between the wars, joined the Post Department as a telegraph message boy in 1906.[46] After training in Morse code to the point where he could send and receive 20 words per minute, he became a wireless operator at the Wellington station in 1912 and worked there until he joined the army as a wireless operator. During a 1955 interview he described the thrill of listening for Morse signals from shipping, as well as from the Australian and Pacific stations: 'it felt like we were eavesdropping on the world'.[47] Wireless had an advantage over other media like films and phonographs, in that it was instantaneous and was multi-channelled in a way that wired telegraph was not. Its open-ended and mysterious possibilities made the routine work of government operators especially interesting and exciting.

Most people were not trained to decipher Morse code, and did not have access to the earphones and other technology required to hear wireless. Instead they received and understood wireless through the written word; they read about it in newspapers and magazines. Its sounds might

This photo of the interior of the wireless station at the top of Te Ahumairangi Hill (Mt Etako or Tinakori Hill) shows the state-of-the-art equipment used. An operator would spend hours wearing earphones and with pencil and pad at the ready for any incoming messages. The earphones were robust and well padded, designed for comfort. Mt Etako station, Radio Wellington, Sydney Charles, Smith, Photographs of New Zealand, Alexander Turnbull Library, Wellington, Ref: 1/1-020077-G

be incomprehensible but it was portrayed and understood as a marvel of modernity.

Some idea of how wireless and modernity were received by New Zealanders is provided by an article in the magazine *Progress*. Established in 1905, the editorial of the first issue stated that the magazine was aimed at 'an appreciative circle of readers from the intelligent schoolboy to the critical adult' and that it intended to report on 'progress in engineering, processes, inventions, industrial work and economics'.[48] Wireless was exactly the sort of technology to interest its readers, and more than 70 articles on the subject were featured between 1905 and 1910 when *Progress* became an architectural journal. The 2 January 1907 issue carried a long article by Captain Louis E. Walker.[49] Walker was the Australian representative of the Marconi Company and was trying to interest the New Zealand government in purchasing equipment for the proposed network of wireless stations.

From the Governor of Victoria to the Governor of Tasmania:

"Victoria salutes her sister State Tasmania."—TALBOT.

The Prime Minister of the Commonwealth to the people of Tasmania.

"Australia, tirelessly subduing her great distances by rail and wire to-day, enlists the waves of the ether in perfecting the union between her people in Tasmania and upon the mainland."

There can surely be no question as to the desirability of establishing wireless telegraphy stations around the coasts of New Zealand, in fact, it would seem in every respect an ideal country for the adoption of this most useful and up-to-date method of communicating between the shore and ships at sea between the mainland and isolated islands, and between ships at sea, more especially when one considers the large proportion of the inhabitants who are constantly travelling round the coasts in passenger steamers. The additional sense of security when travelling on a vessel equipped with the Marconi system has only to be experienced once to be fully realised, but this feeling of safety and of not being cut off is not merely confined to the passenger, but is also appreciated by all his, or her, friends and relatives on shore.

The Marconi Company has established the first wireless telegraph stations in New Zealand, namely, at the Exhibition at Christchurch, where there is a station installed in the Post Office annexe, which is in daily communication with a station at Islington, demonstrations being given constantly, when he who wishes may see this marvellous invention in actual work. Let us hope that, in its small way, it is the forerunner of a comprehensive scheme for the entire colony.

Thus we bring to a close our cursory history of an invention which has increased the facilities for human intercourse, forged new links between lands separated by the sea, aided journalism, given new data to the meteorologist, provided a safeguard for future geographical explorations, and added to the pleasure, while it has diminished the peril, of ocean travel.

Details of the Marconi System.

By G.A.P.

The arrangements for working the Marconi installation are sure to be much discussed during the next few months, by reason of the negotiations pending between the Government of this country and the Marconi Company for the use of their system here. Another reason is the discussion which has been going on at the recent conference in Berlin, at which the German Government attempted, but failed, to generalise the use of all installations. A third reason is the claim put forward by the Danish inventor, Valdemar Poulsen, for the discovery of a system of tuning wireless messages so as to render them absolutely safe from interruption and discovery of the kind sometimes known to have been experienced in the working of some existing systems of wireless telegraphy. Sir William Preece, it will be remembered, declared a few days ago, having heard the Danish inventor explain his system during a lecture in London, that the same was certain to supersede all known systems. This produced many comments in the New Zealand press, which in their turn obtained from Captain Walker, the representative in Australasia of the Marconi Company, a letter of explanation to the effect that, firstly, Sir William Preece is not an authority on a subject which does not lie within the sphere of his expert experience, and, secondly, that Marconi had tried the system of continuous sparks, which is the foundation of the Poulsen invention, and after many experiments discarded it for his own, the distinctive feature of which is the intermittent sparking system. It is further claimed for the Marconi system that it is capable of being "tuned" or syntonised, as the electricians call it, with just as good results as any that can have been obtained by the Poulsen process, or any other for that matter.

With reference to this point, Mr White, of the Engineer-in-Chief's department of the General Post Office, London, dealing with the Marconi system, says in his little book on Wireless Telegraphy. "It is found that the best syntonic effects are obtained with a comparatively weak coupling. There is a limit, of course, to the extent to which the coupling can be weakened. If carried too far, then the current of energy supplied to the aerial circuit, and radiated therefrom, would be insufficient to produce effects at any great distance. A strong coupling would correspond to the case in horology where the balance-wheel was connected as closely as possible to the driving system, the earlier forms of clock mechanism being of this character. Both in horology and in radio-telegraphy it is good policy to have the connection between the oscillating system and the source of energy as far apart as possible." For these reasons we give subjoined an illustration of one of the transmitting and receiving processes of the Marconi system for comparatively limited distances.

The Text Book of the International School publishes the following with the remark that the arrangement of the transmitting and receiving apparatus patented by Marconi, and said to be used by him, is shown in the figure.

TRANSMITTING APPARATUS.

The essential part of the transmitting apparatus is an induction, or Ruhmkorff, coil, as it is commonly called. The primary winding p and the secondary winding s of the Ruhmkorff coil are both wound upon the same iron core, which is here represented, merely for the sake of clearness, as lying between the two coils p and s. The current may be rapidly interrupted by almost any form of interrupter, and a condenser C must be connected across the break $c d$. The condenser reduces the sparking between c and d, and also improves the action of the coil by causing a more sudden interruption of the current that flows from the battery B through the primary p. Both d and c, where they come in contact with each other, are tipped with platinum to better resist corrosion and fusion. Marconi says he found it advantageous to rapidly revolve the contact d by means of an electric motor of some kind geared to the wheel h. By this means the platinum contact surfaces on d and c are kept smooth, and any tendency to stick is removed, and also they last longer.

When the key K is closed a constant stream of sparks will pass between the large centre sphere and the two smaller spheres, one on each side. The

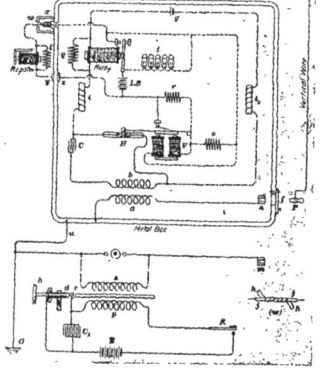

DIAGRAM OF THE MARCONI TRANSMITTING AND RECEIVING APPARATUS

total air gap usually varies from ¼ in to 2 ins., but the coil must be powerful enough to give an 8 or 10-inch spark. One of these small spheres is grounded, and the other connected to a long vertical wire.

The current in the oscillator (merely the circuit from the top of the vertical wire through m to the ground G) surges back and forth between 100,000,000 and 200,000,000 times per second each time it does so it charges or discharges the long vertical wire. The charging and discharging currents flow up and down the vertical wire, and consequently produce electro-magnetic waves that are projected out into space, as horizontal circular waves, from every part of the vertical wire. Furthermore, on account of the static disturbances that are produced in the surrounding space between the vertical wire and the surface of the earth, due to the electro-static capacity of the vertical wire, it is probable that so-called electric waves, which vibrate up and down in vertical planes, are also projected out into space.

Since these waves spread out through space in all directions, it is evident that another vertical wire, if not too far distant, will be cut by some of them. The waves that cut the second vertical wire seem to set up oscillating currents that follow it down to the earth.

RECEIVING APPARATUS.

To prevent the oscillations generated at a station from acting on its own coherer and rapidly destroying the same, Marconi encloses all the receiving apparatus, with the exception of the Morse register, in a metal box, and leads the wire connecting to the register through a coil encased in bands of tinfoil, the tinfoil being connected to earth. The box is usually made of iron, merely because it is the cheapest material. The metal need be only 1/26 or 1/16 of an inch thick. The hole at f should be securely closed by a metal door when transmitting. To receive, the door is opened and the plug P inserted in the receptacle n. The current waves that slide or follow down the vertical wire pass through the primary winding a of a step-up *induction coil*, or *transformer*, as it may be called, when they pass through the metal of the box and the wire u to the ground G. The secondary b of this coil is connected in series with a condenser C and a coherer H.

The induction coil or transformer $a b$ should be in tune, or *syntony*, as it is called, with the electrical oscillations transmitted the most appropriate number of turns and the most appropriate size of wire varying with the length of the wave. Marconi says in one of his patents that he obtained the best results (presumably for 10-inch waves) by using a transformer which he duly described. This has been improved upon considerably since the first of the operations of the Marconi Company.

RATE OF WORKING.

By means of the key K the current flowing in the primary coil may be broken up into ordinary Morse signals. This will cause waves to be projected into space according to the Morse code. To be sure, each dot consists of millions of waves, but all waves cease when the key is opened. The key K used by Marconi when in America was not an ordinary telegraph key in the strictest sense, although it was somewhat similar. It had a longer lever (about 14 to 18 inches) pivoted at about its middle but instead of a finger button there was a handle extending upwards about three inches. The key was moved up and down over a wide gap in order to break the spark in the primary circuit when it was opened. This accounts for the fact that the speed of twelve or fifteen words a minute seems to be about the best so far attained, while ten words is a good average speed.*

MULTIPLEX WORKING.

"By means of loose coupling" says Mr White, "Marconi has succeeded in designing an apparatus for what may be termed wireless multiplex working."

RADIATION OF WAVES CONFINED TO CERTAIN DIRECTIONS.

According to a note written in March last by Mr. Marconi, and communicated to the Royal Society a few days later by Dr Fleming, F.R.S., "when a horizontal conductor is substituted for the usual vertical apparatus it receives with maximum efficiency only when the transmitter is situated in the vertical plane of the said horizontal receiving conductor and in such a direction that the end connected to the detector and to the ground is pointing towards the transmitting station. The wireless telegraph station on H M S *Furious* consisted of an ordinary vertical wire about fifty metres in length connected to a suitable spark gap. The station on the ship transmitted at intervals, and the ship followed a course describing an arc of about 180° round Poldhu keeping a distance varying up to sixteen miles. By means of the horizontal wire arrangement the bearing of the ship from Poldhu could be determined at any time by noting on which particular wire or wires the reception of signals was strongest and also by observing which wires were non-receptive."

Here we have the method employed in finding ships at sea for purposes of communication. In this direction Marconi announced his intention of making further experiments.

---

* Since the above was written the average speed has been increased to thirty words.—ED. PROGRESS.

Cut this out and return with Five Shillings.

The Editor, "Progress,"
Progress Buildings, Cuba Street,
Wellington.

Please place my name on Subscribers' List for one copy of "Progress" each month for twelve months from next issue.

I enclose Postal Note for Five Shillings in payment of Subscription.

Name ....

Address ....

He had already arranged a successful display of the Marconi system at the 1906–07 Christchurch Exhibition, and this was operating when his article was published.[50]

Walker's article gave a history of wireless development to date, or at least of the triumphs of the Marconi Company in this field. Complementing his puff piece was a more technical article that also included a circuit diagram for the Marconi device.[51] This is an example of the sort of literature used by amateurs to make their own wireless sets. It would take more information than this to build a set that worked, but the piece laid out the basic principles of the hardware and how it would be wired.

Similar guides and diagrams were found in magazines such as *Popular Mechanics*, and newspapers also published articles that explained the technical aspects of wireless in some detail.[52] Manuals, handbooks and the equivalents of present-day 'How to' guides were readily available from bookshops or could be bought from overseas. Oliver Lodge's *Signalling Through Space Without Wires* (London, 1900), George Pierce's *Principles of Wireless Telegraphy* (New York, 1910) and C.C.F. Monckton's *Radio-Telegraphy* (London, 1908) were the sorts of works consulted by New Zealand radio enthusiasts for circuit diagrams, equations and general technical advice.

An introduction to wireless that had broader appeal was the radio station set up by the Marconi Company at the Christchurch Exhibition. This was the first public display of wireless in New Zealand, and the regular demonstrations sparked a great deal of interest.[53] The exhibition itself was an important moment in that it marked the country's transition from Colony to Dominion status. Just as the world was being brought to New Zealand in the displays from other countries participating in the exhibition, New Zealand was displayed to the world, and to its own citizens, as a prosperous, advanced country.[54] The new dominion was not going to be an isolated and peripheral island, remote from the world centres of North America and Europe. And new technologies, including wireless telegraphy, were going to play an important part in ensuring it was a modern, connected society.

• • •

Left: The information about wireless in this article would have interested many of the magazine's general readers, while the finer points of the circuit diagram would have been devoured by the country's 'boy heroes' of wireless. Progress, *2 January 1907, 94*

Too Much Easter: Intoxicated Citizen (fondly embracing a Lambton Quay telegraph post: 'Don't talk – hic – to me – hic – of new inventions. What'll I do – hic – when they go it – hic – for this newfangled – hic – wireless telegraphy?'

The drunk had nothing to fear. Wireless may have replaced telegrams but the poles were still needed for the telephone lines as that communication technology spread throughout New Zealand. New Zealand Free Lance, 6 April 1907, 15

Apart from the Christchurch Exhibition, most New Zealanders would have received radio through the pages of newspapers and popular magazines rather than technical manuals and specialised journals. Local media had been connected to the global network of telegraph lines since 1876, and regularly reported on Guglielmo Marconi's experiments with wireless from 1897.[55] The world's news, fashions, fads and technological developments, including radio, were part of the daily life of the country's people.[56] To understand wireless and other technologies was to be familiar with the progress of science, the annihilation of time and space, the shrinking of the world, the increasing, exciting and nerve-tingling acceleration of the speed of modern life. To know about and maybe use these things was to be part of modernity, and New Zealand was a country that was 'born modern'. It entered and participated in the modern world of technology, mass media, and consumption of industrially produced global culture from the establishment of the country as a nation-state.[57]

Part of being modern was a form of nonchalance in the face of seemingly miraculous technologies such as wireless communication. Telegraphs, telephones, recorded sound and films were all accepted parts of New Zealand life by the time wireless was demonstrated at the 1906–07 exhibition. Cartoons and jokes, along with sober technical accounts and demonstrations, had played important roles in the normalisation of this technology. To get the joke meant that you got the technology, even if you

SIR JOE IN A HURRY
First Tramp (as something dashes past): 'Whew! What was that? Was it a streak of lightning or a wireless telegraph message?'

Second Tramp (who has been reading the election news): 'Well, I rather guess it was Sir Joseph Ward on his election tour.' New Zealand Free Lance, *7 November 1908, 12*

did not actually have it as such. In a cartoon from the period, a drunken man woozily clinging to a telegraph pole on Lambton Quay wonders what he will do for support when wireless telegraphy renders wired telegraph poles obsolete. It plays on the audience's understanding that wireless telegraphy is literally telegraphy without wires – an important aspect of the new technology that people knew even before they saw it. Such familiarity helped demystify the technology itself.

A second characteristic of wireless was its speed. Another cartoon drives this point home with little or no subtlety. The speeding form that has just passed by has confused a tramp, who wonders if it was lightning or a 'wireless telegram'. His companion reassures him that it was only the prime minister on the campaign trail. The message is that wireless covers large distances very, very quickly. Readers would have understood, too, the relevance of a reference to Joseph Ward. Ward was Postmaster-General in 1905–06 when he toured Europe and the US and attended many demonstrations by various wireless companies. He became familiar with the technology and understood its strategic and commercial possibilities. After becoming premier in the 1908 election, he pushed along the development of official wireless communications in New Zealand.[58]

Although wireless was celebrated as a new technology, for some it was also imbued with magical powers, and was associated with spiritualism and psychic experiences.[59] The international interest in spiritualism was carried into New Zealand as numerous mediums and speakers toured the country. There were also homegrown attempts to communicate with the dead.[60] Audio technologies such as the telephone and the phonograph had detached the voice from the body, conquered space and time, and provided metaphors (and sometimes mechanisms) for 'communication' with the dead.

New Zealand's spiritualists, devotees of telepathy, ghost hunters and even astrologers were quick to seize on wireless telegraphy as a mode of other-worldly communication. Hence a 1903 article described telepathy as 'the wireless telegraphy of the mind'.[61] In 1906 Nelson astrologer Joseph Taylor compared finding an individual's 'cosmic key' to wireless tuning.[62] The theosophist Annie Besant, when interviewed during a 1908 lecture tour of New Zealand, spoke of the power of thought as a natural force like wireless telegraphy and 'most potent in its effects when rightly understood and directed'.[63] Wireless was not being understood here as a mystery, but rather as a metaphor that could explain mysterious ways of communicating such as telepathy or contacting the dead.

The explanatory power of the wireless model was based on the physical nature of the signals: they are invisible but natural in the sense that they obey physical laws and can be manipulated for communication purposes. Spiritualists thought of consciousness as a form of energy that survived the death of the body, and was a natural part of the universe just as radio waves are. The vibrations of the living and the dead could thus be tuned into in the same way that a wireless tuned into signals. This is what Annie Besant meant when she compared thought to wireless signals.[64] The technologies that separated a person's voice from their presence, such as the telegraph, telephone and phonograph, provided models for ways of thinking about disembodied consciousness that bypassed the body. Wireless took this idea even further, as its invisible waves and signals seemed to free communication from space and time.[65]

• • •

For most New Zealanders, however, tuning into wireless meant tuning into this world rather than another. While very few may have physically heard a wireless set before the early 1920s, most had heard of the technology and had ideas about its potential. It was a sign of an advanced and technological age that had already delivered recorded sound and moving pictures. A sailor might see wireless as a navigation aid, while a spiritualist might see it as a model that helped explain telepathy or clairvoyance. Both were attempts at imagining the effects of the ability to transmit sounds over distance. What was important about the effects of wireless on New Zealand was not its sounds but the echoes of these sounds in newspapers, magazines and books. Wireless opened up new ways of receiving and imagining the world. When war was declared in August 1914, the first military action undertaken by New Zealand was the capture of the German radio station at Samoa. Modern technology was at the heart of this modern war.

Chapter 7

MILITARY RADIO

•••

When war broke out in 1914, New Zealand's network of government-operated wireless stations extended from Awanui, near Kaitaia, to Awarua just south of Invercargill, and included a facility at Waitangi on the Chatham Islands. This network began monitoring and intercepting German wireless transmissions within days. Intercepts were passed on to the Admiralty in London and the Australian Commonwealth Naval Board in Melbourne.[1] New Zealand's wireless listening was for the security of the state and the safety of the armed forces. Just how important this listening could be was soon established.

Clive Drummond had joined the Post and Telegraph Department as a telegraph messenger in 1906 after leaving high school. By 1908 he was being trained in the mysteries of Morse and the new technology of wireless at the department's school at Ōamaru. By 1912 he was a fully fledged P&T wireless operator stationed at Wellington's Tinakori Hill Morse Wireless Station.[2] When on 23 September 1914 New Zealand troops around the country struck their camps and boarded transport ships at Auckland, Wellington, Christchurch and Dunedin, it was expected they would form a convoy in the Tasman Sea and then join up with Australian forces.[3] That night Drummond was on duty at Tinakori Hill and intercepted a message between two German warships in the Pacific which indicated that the supposedly secret sailing of New Zealand's armed forces was in fact known to the German naval command and a squadron was preparing to intercept the hapless transports.[4] The New Zealand transports were recalled to port and a stronger escort was arranged. The Main Body finally set sail on 16 October 1914 for Egypt and, ultimately, Gallipoli. The strategic power of wireless in the new age of global and modern technological warfare had been decisively confirmed.

New Zealand wireless was militarised between 1914 and 1918. No civilian wireless activity was allowed and such activity was heavily punished when it was detected.⁵ But the war years increased the number of New Zealanders who used wireless. There was a growing demand for wireless operators from the armed forces and the merchant navy, and new civilian operators were required to replace those who joined up or were conscripted. New Zealand's World War One wireless operators met their overseas counterparts on active duty and caught up with all the latest developments in technology and techniques. The knowledge, skills and equipment (acquired in various ways) that they brought back to New Zealand after the war were crucial for the development during the early 1920s of radio broadcasting – that is, broadcasting in the sense of a radio station transmitting sounds like words and music to anybody in range who could tune in.

...

The strategic importance of wireless for modern warfare was emphasised by the first military action by New Zealand in 1914: the seizure of the German wireless base at Samoa.⁶ Germany had annexed Samoa in 1900 in order to further its economic and strategic interests in the Pacific.⁷ It had also established a large wireless station on the main island of Upolu in the hills behind Apia, and this became an important link in the global communication network that Germany maintained between Berlin and its colonies, fleets and garrisons around the world. The British government wanted the station neutralised, and the New Zealand government had long aspired to control Samoa. The British suggestion that the island be invaded and the station captured was enthusiastically received in New Zealand, and a 1400-strong force was quickly despatched. Samoa was captured by New Zealand with no opposition on 29 August 1914.⁸ New Zealand legend has it that this was the first piece of German territory captured during the war, but this is not the case. French and British forces had captured the German colony of Togoland in West Africa on 26 August 1914.⁹

The Apia wireless station had been sabotaged, but it was soon repaired and became part of a Pacific wireless network that included Suva and the New Zealand stations. These were used for monitoring both enemy and neutral radio traffic. The operators spent long hours listening for suspicious signals among the friendly traffic and the ever-present interference, atmospheric noises and static. The duties of the listeners at Apia and the other stations were laid out in a 1916 cable from the governor of New Zealand which enjoined them to keep a continuous watch for '… suspicious radio signals: enemy signals possibly not in code but en clair in the form of messages between neutral stations'.[10] The watchers (or listeners in this case) had to 'record all work overheard and scrutinise [it] for hidden meaning'. They were to 'communicate suspicious messages with particulars to call signs, wave lengths and any other information likely to be useful to "Naval" Wellington'. These details would then be further scrutinised and passed up the chain of command to form part of Britain's global strategic planning.

This was often tedious work. Army routines filled in much of the time, but there were few diversions. Time weighed heavily and many of the soldiers began to feel isolated from the world.[11] One way of avoiding boredom was through writing and publishing a small magazine. 'Troop' newspapers of this kind were put together by soldiers on both sides, and in the case of New Zealanders were produced on ships, in camps, at hospitals and in all the theatres of operation.[12] The Samoa garrison produced five issues of its magazine, *The Pull Thro'*.[13] Among the usual jokes, complaints, amateur poetry and sentimental evocations of home, one anonymous writer lamented the isolation and also pointed to the wireless as a source of news and information, even if it was fragmentary and incomplete:

> The arrival of the mail from home is a big event in our lives now-a-days. Cut off as we are from the outside world, with only meagre wireless details of happenings in the cockpit of Europe, filtering slowly through, even weeks' old newspapers are priceless treasures, while letters – well every member of the force knows full well what his feelings are when a letter from home is received. The scorching sun, the monotonous routine of garrison life are forgotten.[14]

The wireless station at Apia with its nearly 120-metre mast was a major piece of state-of-the-art communication technology. The machinery was complex and needed well-trained technicians to keep it in working order. A description in *The Pull Thro'* captures some of the awe this technology could inspire. The station is described as 'the sanctum of the chief wizard of this Island' who controls 'five hundred thousand electrical devils'.[15] The operation of the station involves 'bright blue-green sparks … the sound of a rushing, roaring wind … violet flames'. But the more prosaic description that follows this flight of fancy is equally impressive with its account of the cost of the station as a 'quarter of a million pounds' and the 'tower weighing 40 tons'. The writer goes on to justify the expense and trouble of capturing the station (as well as the boredom of garrison duty) by pointing to its long-term value. It is 'the nucleus of a new colony, which in time to come will

The German radio station in Samoa was a strategic asset that was quickly seized by New Zealand in 1914 as part of the Dominion's war effort and imperial ambitions.
Archives New Zealand, ACHK 16604, G49/10, R19162313

be the stepping stone from one continent to another, as it is at present the half way "talking house" of the Pacific'. Here magic, modern technology and imperialism combine to explain the importance of wireless. The garrison was relieved in 1915, although a small force remained on Samoa for the rest of the war. The wireless station was a valuable strategic asset that could not be left unguarded.

The war generated urgent demands for more wireless operators than had been produced during peace time. Merchant ships, navy ships, the army and the many transports that conveyed New Zealand troops around the world all needed qualified and competent operators to maintain communications. This increased demand could not be met by the postal service alone, and the Dominion College of Radio-telegraphy Ltd appeared to plug the gap. This organisation was established in 1915; it promised to 'qualify students for [the] International Wireless Operator's Certificate' and offered 'practical tuition by qualified experts'. Wireless was 'the profession of today and of tomorrow'.[16] The college advertised heavily and promoted correspondence courses as well as classroom instruction. It stressed the importance of wireless in modern war and how such skills provided opportunities for service and advancement. Here was a way to receive wireless that might not only help win the war but also advance personal careers.

Equipment and teachers for the Dominion College came from the Amalgamated Wireless Company of Australasia Limited.[17] Incorporated as a private company in 1917, the college eventually had premises in Auckland,

The Dominion College provided a useful education for both peace and war. New Zealand Observer, *28 October 1916*, 23

Wellington and Christchurch.[18] Prime Minister Joseph Ward opened a new Auckland head office in 1918 and spoke glowingly of the contribution the college had made to the war effort. In one year it had enrolled 250 students, 89 of whom had gone on to military wireless work; four had died on service. Ward stressed the importance of wireless, and noted that: 'In these times of war it had been found necessary to exercise the greatest care that only persons whose loyalty was undoubted should be instructed in wireless.'[19] Applicants were vetted by the Postal and Telegraph Department and '[t]here had been several cases in which it had been found advisable to decline permission'. It was dangerous to let the wrong ears listen in to military radio during war time.

The growing importance of wireless for New Zealand's communication networks extended beyond the war. When George McDonald, managing director of the college, was interviewed before the end of hostilities in 1918, he spoke of the heavy demand for wireless operators and stated: 'After the war there will be many fine positions in this profession, and lads who intend going in for "wireless" would be well advised to learn something about it.'[20] But in the meantime, the shortage of trained operators had serious implications for the war effort.

In June 1918 a Military Review Board heard evidence that a troop transport departure had been delayed due to a lack of qualified wireless operators.[21] Special permission had to be sought to allow a boy who was not yet 17 to act as the vessel's wireless operator, as no one else could be found. The board also heard that the Defence Department had been required to borrow uncertified wireless operators from the college on several occasions to ensure that transports had sailed on schedule. The Post and Telegraph Department and the three schools owned by the Dominion College simply could not train enough operators to keep up with the demand.

The college continued its training programmes throughout the war and into peace time, but the demand for operators declined after the Armistice when the trained operators returned home. This was bad news for the college, which went into liquidation in 1920 and disgruntled students resorted to the courts to try to recover their fees.[22] New technology companies are still often victims of market churn.

• • •

The majority of New Zealand's wartime wireless operators were dispersed around individual units or on ships. Only one unit was formed to specifically use the new technology. The New Zealand Wireless Troop numbered only 179 personnel and operated in the Mesopotamian and Persian campaigns from 1916 until early 1918, when it was sent to France and merged with the New Zealand Divisional Signallers.[23]

The Mesopotamian Campaign was regarded then as a sideshow and it is still something of a footnote for many modern histories of World War One. However, it was a major conflict and was in many ways a template for a number of current wars. It involved imperial ambitions and causes. It used the latest communication technology: in this case radio. The armies involved were international ones, composed of soldiers whose governments were committed to trans-national alliances. State-of-the-art intelligence gathering and code breaking were vital parts of the Mesopotamian Campaign. Local religious, tribal and national interests were invoked and provoked in attempts to secure cooperation by imperial powers. Islam was an important factor in shaping alliances and motivations on both sides. The bulk of the troops on the British side were Indian and many of them were Muslim and were fighting their co-religionists. All up, about half a million people were involved in the armies on both sides.[24]

On 28 December 1915 the Indian government asked the New Zealand government to supply the personnel for a Pack Wireless Troop to be used in Mesopotamia. The New Zealand government agreed, and the first signals unit to specialise in the new technology of wireless was the result.[25] The men of the Wireless Troop were from the New Zealand Post and Telegraph Department, and began training at Trentham military camp, near Wellington, in early 1916. There they used an old radio set from Wellington's General Post Office and became accustomed to military life.[26] Initially there were 12 experienced wireless operators and 18 qualified telegraphists. Wireless operation was a specialised skill demanding varied technical abilities and a fluent grasp of Morse code. Most of the men involved knew each other from working in the Post Office; many of them had been interested in wireless from school days, and had experimented with making their own sets. They had in common specialised skills, technical interests and employment histories. This made them a close-knit unit.

The Wireless Troop, along with 1800 other soldiers, left New Zealand on the SS *Willochra* on 4 March 1916 as part of the Tenth Reinforcements. The journey took them through Colombo in Sri Lanka and eventually to Bombay.[27] Like the men of the Samoan garrison, those aboard published a journal. Called the *Willochra Tatler*, this magazine lightheartedly promised '[t]he unvarnished truth of some companies of the 10th reinforcements N.Z.E.F.' in the form of 'articles on sports, company notes, musical notes, verse, the voyage and personal powers etc etc etc'.[28] It also contained an embarkation roll which listed all the members of the 10th Reinforcements, including the Wireless Troop.[29] It featured an article about the Wireless Troop written in a gently ribbing tone which reflected the atmosphere of the temporary communities the troopships formed. This praised the 'Sparkies' for their geniality and sporting prowess before concluding with best wishes and describing them as 'jolly good fellows'.[30]

The Wireless Troop suffered its first casualty during the voyage when Sapper Llewellyn McMillan died of fever. He was buried at sea in an 'impressive service' during which the ship was stopped and the troops were drawn up in parade formation.[31]

The sporting prowess of the so-called 'Sparkies' noted in the *Willochra Tatler* cropped up again in a letter describing the group's journey to Basra. During a stopover in Bombay, the Wireless Troop played rugby against the local YMCA, 'described as the best team in the city', and won 11–0.[32] Rugby and war were and remain major components of both New Zealand masculinity and the national imagination, and even though the Sparkies were boffinish types working with highly sophisticated technology, they were also making it clear that they were 'good keen blokes'.

The New Zealanders finally arrived at Basra on 16 April 1916 and spent some time acclimatising and training. In July they were amalgamated with the Australian forces and formed the Anzac Wireless Squadron. This was a relatively quiet phase in the Mesopotamian Campaign. The Indian–Anglo forces were being reinforced and rebuilt after the capture of Kut by the Turkish army. Nevertheless, living conditions were difficult. The New Zealanders found the climate oppressive and fell prey to disease. By October 1916, 40 of the 62 were in hospital. When much-needed reinforcements arrived that month, half of them fell ill within weeks.[33] In the end, 13 of the 179 soldiers who served with the Wireless Troop died, all of them from diseases, mainly dysentery.[34]

The letters written by the Wireless Troop soldiers all mention the heat and general discomfort. Sergeant Major Richard Croucher of the Wireless Troop had been part of the force that captured the Samoa wireless station and eventually joined the Royal Flying Corps.[35] Croucher described Mesopotamia as 'an inhospitable land'. His letter is worth quoting as an example of how the New Zealand troops reacted to their surroundings:

> Expected great things of this Mesopotamia and was disgusted to find that the land, so famed and glorified in Biblical history, as being a country wonderfully productive, redolent of milk and honey, civilised beyond the ordinary, the homeland of Adam and Eve, Sinbad the Sailor, and countless others, was in actuality a barren, inhospitable land, cruel and pitiless. The summer temperatures were simply awful; just imagine 125 degrees or 130 degrees in the shade, at times, and seldom below 115 degrees [46.1°C]. I was quite disappointed in the wonderful Garden of Eden which is only a degree better than the surrounding, barren, sun-baked, mud desert, inasmuch that it grows a few acres of date palms, fig, lime, and apple trees. And the apples, my goodness, they were the limit, quite devoid of flavour, and the colour of grass and hardly the size of a golf-ball. Our forces are fighting Mother Nature, a force more formidable than the Turk as our casualties show.[36]

Home comforts were few and far between, and greatly welcomed when they arrived. Some of the troop felt they were neglected. The writer of one letter published in several newspapers wondered if the various patriotic and gift associations had forgotten them: 'Though small in numbers compared with the other expeditionary forces we are just in the place where comforts are most needed because nothing can be procured here.'[37]

But the Wireless Troop soldiers were fortunate in some ways. They received half of their normal Postal and Telegraph wages on top of their army pay, although this was usually sent to their parents or wives.[38] The P&T also organised social events such as dances to raise funds and supplies specifically for the Sparkies in Mesopotamia.[39]

To break the boredom and stay in training during their first few months in Mesopotamia, the Wireless Troop began taking down all the wireless messages they could pick up. These included Turkish and Russian messages, as well as press wireless from the Germans, French, English and Indians. The military messages were in cipher so could not be understood, but the wireless operators pinpointed the locations of the Turkish and other stations and could recognise their particular calls, such as DAS for Damascus and SMR for Samarra, SBA for Mosul and so on. There were at least a dozen Turkish-operated wireless stations in Mesopotamia, so there was plenty of material for the Wireless Troop to gather.[40]

The British War Office sent linguist Gerard Clauson and archaeologist Reginald Thompson to Mesopotamia as codebreakers. Within one day of arriving they had broken the Turkish ciphers. According to one of the troop: 'Despite daily changes and enciphering of a most complicated kind, every enemy message arrived at Intelligence as surely and certainly as if it had been addressed to them.'[41] This information was vital for the Indian–Anglo army that was being prepared to advance on Baghdad. Turkish troop positions, movements, supplies and strengths were all clearly known through the wireless intercepts, and information brought in by spies could be confirmed or rejected. General Maude, commander of the Indian–Anglo army in Mesopotamia, stated that 'the messages agreed so exactly with previous information that there was only a discrepancy of 2 guns and 17 men over the whole Turkish North-Eastern Front'.[42]

Intelligence about allies is just as important as that about foes, and much effort was made to listen in on the Russian messages, although this was harder work as it involved a new Morse code to take in the extra letters in the Cyrillic alphabet. Before the war Russia had been seen as a major threat to India and this did not change even when both sides were fighting Germany.

The Wireless Troop was closely involved in the advance on Baghdad that began in 1916 and among the first troops to enter the city on 11 March 1917. General Maude understood the value of wireless and demanded regular reports from any units on patrol or acting as units detached from the main army. The wireless pack sets used for mobile work were portable and rugged, and could be set up in seven minutes or so. Three of these sets would accompany a mobile column. One would be set up as required while the column continued to advance. When the transmissions were complete, the wireless operators would pack up and gallop to catch up with the column, which may well have deployed another set in the meantime. The range of

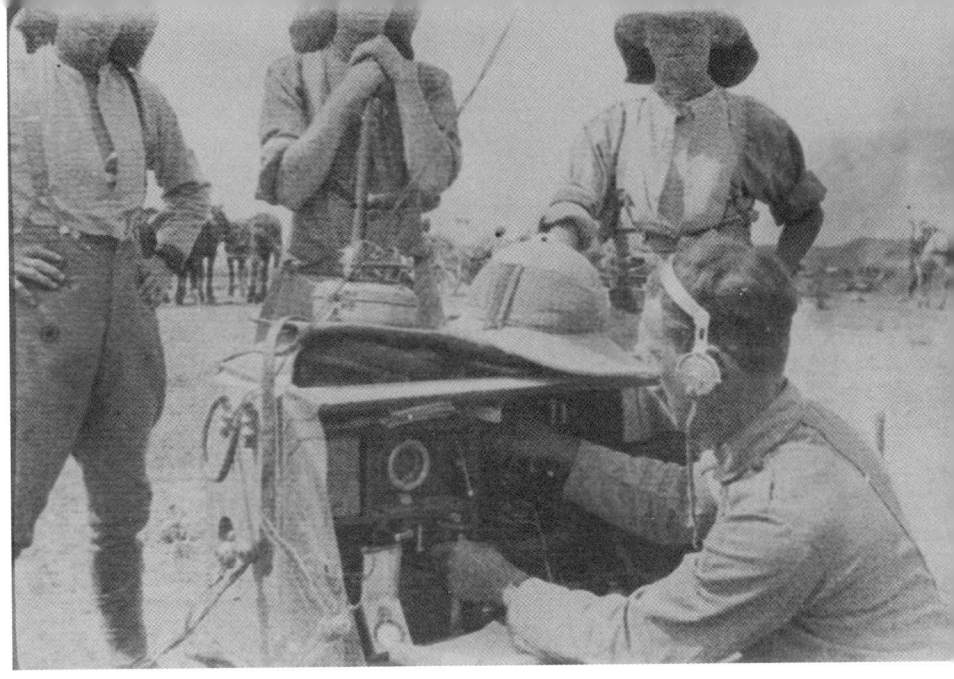

Operating a pack set in Mesopotamia. The generator for the set was built on a framework that would sit on a horse or camel. These machines were hard to use but highly mobile. Note the earphones. Hocken Collections Uare Taoka o Hakena, P1993-024-022b

these sets was up to 60 kilometres, and with this leapfrogging pattern a mobile column could maintain continuous contact with headquarters.[43]

In Mesopotamia and Persia the open and mobile nature of the campaigning meant that communication systems which did not depend on cables were highly effective. But this pioneering radio technology was often not easy listening. Intense concentration might be required to hear the faint wireless signals, and a great deal of ingenuity was needed to make the equipment function in the harsh conditions. In one instance a sergeant dried his set with a blowlamp during torrential rain. When the sound of horses' hooves in the thick mud drowned out the faint Morse sounds in his earphones, troops cleared the surrounding area and the operator crouched in his tent with thick coats draped over his head as impromptu sound proofing. Each part of the ciphered message was sent nine times and this was a slow process. The message got through but the sergeant 'finished in a condition approaching collapse'.[44]

In May 1918 the Wireless Troop was sent to France and incorporated into the New Zealand Divisional Signal Company. Many of the soldiers became part of the forces that occupied Germany and did not return home until

In March 1919 these New Zealand soldiers were stationed in the German town of Mulheim and getting ready for demobilisation. The illustration on the blackboard is captioned 'Transmitting circuits'. As the Dominion College had promised in its advertising, wireless was the coming thing for ambitious young men. Royal New Zealand Returned and Services Association, Alexander Turnbull Library, Wellington, Ref: 1/1-002105-G

late in 1919. During this time they caught up with the latest developments in radio technology and took advantage of the educational opportunities offered by the New Zealand army as part of the process of easing soldiers back into civilian life.[45]

The New Zealanders involved in wireless work overseas were able to learn about new technologies and developments from the contacts they made with other Allied operators, as well from as the education provided by the army. They sometimes acquired these new technologies by fair means or foul. Cob Smith, who served in the Wireless Troop, later recalled: 'We all learned much from our wireless experience in France and returned full of energy for fresh experiments. The valve had been invented – we had used them in France – and most of us managed to stow a few in our inside pockets to take back with us.'[46]

The development of thermionic valves allowed the broadcasting of sounds beyond the tapping of Morse. Speech and music could now be reliably sent and received over large distances. L.E. Strachan, a pre-war radio enthusiast, was at Lyttelton wharf in late 1918 when an associate returned from service as a ship's wireless operator. The officer leaned over the ship's rail and shouted to Strachan, 'Wait till you hear these vacuum tubes. You can play tunes on them.'[47] Local operators also acquired the valves from mail steamers whose wireless technicians would sell them at inflated prices.[48]

With this new technology and the skills acquired in the armed forces, a number of ex-servicemen went on to begin broadcast radio in New Zealand during the early 1920s. Others took a more circuitous route. One of these was young Eric Battershill, whom we met in the previous chapter after he fell foul of the law with his unauthorised wireless experiments. Battershill seems to have been gifted with initiative and versatility. When war broke out in 1914 the 16-year-old tried to enlist but was told he was too young; besides, he was viewed as unreliable due to his wireless activities. Undaunted, he worked his way to Australia on a ship and enlisted there. After joining the Australian motor transport and serving as a driver, wireless operator and intelligence officer, he was sent to France and eventually flew with the Australian Flying Squadron. When the war ended he went back to Australia and worked in a jazz band, taking the stage name of Eric Dare. As radio broadcasting began to take off during the early 1920s, Battershill found a technical research job with a Sydney company that owned a radio station, and eventually became a station manager.[49] He was tinkering with wireless again, only this time he was being paid for it rather than being summonsed.

Clive Drummond had a more straightforward path into radio. Drummond, who had saved the NZEF from German warships in 1914, served with the Wireless Troop in Mesopotamia where he 'saw many places of interest and had some rather unique experiences' – a masterpiece of understatement considering the Troop's adventures.[50] On his return to New Zealand he resumed his work at the station on Tinakori Hill, and in 1922 became involved with a group of Wellington wireless enthusiasts who were experimenting with broadcast radio. It was the beginning of a long and illustrious career.[51]

• • •

Broadcast radio in New Zealand began during the early 1920s when pioneers across New Zealand used radio to transmit voice and music to anyone who could receive their signals.[52] The stringent regulations controlling radio amateurs that had been in force during the war were slackening, and in 1921 the P&T Department was allowed to issue transmitting permits.⁵³ Small groups such as the one Drummond joined were established from Gisborne to Dunedin to Auckland.⁵⁴

The 1921 broadcasts by Professor Robert Jack of Otago University were the stimulus to the other groups around the country. Beginning on 17 January 1921, and interrupted by the university holidays, Jack broadcast across the South Island on Wednesdays and Sundays between the hours of 8pm and 10pm. The broadcasts began with a buzzer for several minutes to allow listeners to tune in. He then transmitted spoken voice and music, both live and recorded: the latter came from a gramophone with a microphone held up to it.⁵⁵ There were numerous breaks in both transmission and reception, and interference of all sorts from other radios and atmospheric conditions. These broadcasts were experiments, during which Jack would describe the equipment he was using and what changes he might be making. He would also ask individuals he had previously contacted to call in and report on various technical aspects of the signal and reception. His listeners were other wireless enthusiasts and experimenters.⁵⁶ Many of these went on to apply for permits and began their own experiments with the new medium. By the end of 1922 there were stations in the four main cities broadcasting regularly, and several others around the country operating on a more sporadic basis.⁵⁷ Radio broadcasting was established, and beginning to attract listeners who were not also broadcasters.

The equipment needed to receive radio was bulky and inelegant. It involved large batteries that were prone to leakage. A tall mast was required to receive signals as well. The equipment had to be made, and needed considerable patience and skill to operate. Radio sets had to be domesticated, just as the record player had been.

But if radio was heard by a small number of people during the early 1920s, the medium was in the air in the sense that it was part of popular culture. Newspapers carried numerous stories about radio and its achievements and possibilities, along with technical information. Regular columns about radio began appearing in New Zealand's newspapers in 1923. The earliest was 'Wireless News', under the by-line 'Electra', which debuted in Christchurch's *Press* in January 1923. This first column was a sturdy diatribe on the absolute need for aerials to be solidly earthed, and included a bibliography of essential reading for any radio enthusiast.⁵⁸ The *Auckland Star* began 'The Magic Spark' by 'Phonos' in June. Phonos proposed to discuss issues of interest to the wireless amateur rather than just give technical advice. The first column

This radio set from 1923 has been built from assorted components and needs large batteries. It would probably not be welcome in the average house – radio had yet to be domesticated. Hocken Collections Uare Taoka o Hakena, E2014/26Filename2383_01_017A

was an optimistic survey of radio activity and prospects around the world and in New Zealand. Phonos concluded with the observation that radio is 'a fascinating and instructive hobby that may lead to more in the future'.[59] Other regular newspaper columns about radio followed under such by-lines as 'Thermion', 'Magna Vox' (the name of a loudspeaker as well as an obvious pun) and 'Rheostat'.[60] These not only gave good advice on technical matters but also discussed issues about radio governance and content, and kept people up to date with new developments. For those who had not yet experienced radio, the columns gave the medium a sense of boundless possibility.

Those who did have radio sets during the early 1920s heard a mixture of speech and music. Much of the music was played live. The stations broadcast for a few hours a day and only several days a week, but they were regular broadcasts, and listeners expected to tune in and hear the transmissions at appointed times. It was the beginning of a major change in domestic life. Home-delivered newspapers also arrived at more or less the same time each day but could be read at any time. Radio broadcasts gave people the same experience at the same time, and this had profound political, social and

THE NEXT ELECTION:
The Prospective Candidate addresses the Constituents by radio.

This cartoon in the British magazine London Opinion *emphasised the mass audience nature of radio and speculated on how it might affect the art of politics. Faceless crowds would be addressed en masse by their masters' disembodied voices ...*

New Zealand Herald, 2 December 1922, 3

cultural effects as the age of mass politics and mass movements developed during the 1920s.

However, in New Zealand, as wireless telegraphy became radio broadcasting during the early 1920s it was satisfying enough to know that anyone was listening to the scattered transmissions. On 3 January 1925 the All Blacks played England at Twickenham, the climatic match of their British 1924–25 tour. The All Blacks won 17–11. The news was instantly telegraphed to Auckland and telephoned right away to Clive Drummond in his Wellington studio. He announced the result at 2.44am New Zealand time, and it was repeated intermittently throughout the morning until 10am. The station received over 60 letters during the following days from listeners who had heard the result on the radio.[61]

By 1925 New Zealand radio had broadcasters and listeners, but what it lacked was systematic and organised funding, administration and governance. This developed from 1925 as the government exercised more and more control. As the medium became more structured, predictable and bureaucratised, so too did radio listening as the rhythms and routines of daily life became enmeshed in the rhythms and routines of broadcasting.

Chapter 8

HEARING RADIO

•••

Gwen Shepherd wore a gown of ivory georgette when she married Bruce Stennett at Wellington's St Paul's Pro-Cathedral, now known as Old St Paul's, on Wednesday 29 January 1930. Gwen had wanted a small, private ceremony, but around 2000 people were waiting in the streets outside St Paul's and all the seats inside were occupied. As Gwen arrived, the crowd cheered loudly and children were lifted by parents to get a better view. There were radio microphones inside the church, where Archdeacon Innes Jones officiated, and Clive Drummond, now a well-known 2YA announcer, led a live broadcast of the ceremony, even though his station was normally off-air on Wednesdays. Outside the church after the ceremony, photographers were snapping away as quickly as they could, and the press of the crowd made it difficult for the bridal car to get clear.[1] The fine weather allowed the reception to be held in the garden of the Shepherd family's home in Moxham Avenue, Kilbirnie. Speeches were made and broadcast by the 2YA microphones placed at the head table. Gwen spoke and told her listeners to do a good deed every day and be kind to animals. She then said a final farewell.[2]

This was hardly the stuff of a typical wedding speech, but this was a special occasion. Gwen Shepherd was known to thousands of New Zealanders, especially children, as Aunt Gwen, and had been one of the hosts of 2YA's 'Children's Hour' since 1927. Along with the other Uncles, Aunts, Cousins and Big Brothers, Aunt Gwen told stories, presented musical items and introduced talks for young radio listeners. Her wedding marked the end of her tenure at 2YA.[3] The news of her resignation in December 1929 caused a flood of letters, postcards and parcels. Children wrote of their sorrow at Gwen's departure. Some women wrote to say how her show had cheered both them and their children. One riddle sent to 2YA asked, '[W]ho is the most unpopular man in New Zealand? Answer – Uncle Bruce,

Aunt Gwen of Radio 2YA at her wedding reception held in her parents' garden. The microphone was ever present at the wedding and the reception, a reminder that she became famous through radio. This was New Zealand's first 'celebrity wedding' to be broadcast live. Miss Gwen Shephard on the occasion of her marriage to Mr B.H. Stennett, Alexander Turnbull Library, Wellington, Ref: 1/1-032983-F

because he is taking Aunt Gwen away.'[4] An estimated 30,000 people tuned in to the broadcast of Gwen's wedding.[5] There were 53,400 radio licences in New Zealand at the time.[6]

As a popular public figure, there was little chance that Gwen's wedding was ever going to be a private affair, and photographs of the occasion appeared in the press. The 2YA logo displayed on the microphone beside Gwen left no doubt about which station had broadcast 'the first radio wedding in New Zealand'.[7] The microphone was used by Clive Drummond to comment on the wedding reception and to also broadcast the speeches, one of which was made by Joseph Davies, the director of 2YA. The broadcast was a technical triumph for the station, and good publicity as well. The presence of a microphone at the wedding made an intimate ceremony a public event. It added 30,000 invisible guests, and demonstrated the possibilities for radio to connect people through sound and imagination.

• • •

The establishment of the Radio Broadcasting Company (RBC) by government regulation in 1925 began the systematic organisation of radio in terms of content and broadcasting schedules. By 1930, when Aunt Gwen was married, there were already 23 radio stations operating in New Zealand.[8] The number of licensed radio receivers in the country in 1925 was 4702; by 1930 the number had grown more than tenfold, to 58,597.[9] Most of New

Zealand's 1.5 million people were within range of at least one radio station.

The four YA stations were operated by the RBC, which was funded by a licence fee paid by those who owned radio sets. The other 19 were known as B stations and were independently owned. These stations maintained a precarious financial existence; they had no access to the funds generated by the licence fees and were forbidden from selling advertising slots. Instead, they relied on donations and the efforts of volunteers while lobbying for regulatory changes that would allow them to make money from advertising. The RBC was succeeded by the Radio Broadcasting Board (1932–36) which also struggled with the B stations. The issue of these stations became a moot point in 1936, when the government took direct control of radio broadcasting with the formation of the New Zealand Broadcasting Service (1936–62). The history of the often stormy relationship between the publicly and privately funded organisations of broadcast radio, and its regulatory background between the wars, has been at the forefront of accounts of New Zealand's radio history.[10] However, this chapter and the next are about those who listened to radio, rather than those who controlled and made it.

The formation of the RBC coincided with developments in radio technology that made the equipment less cumbersome and therefore easier for people to use. Just as the phonograph became more acceptable as a domestic appliance when it began to resemble a piece of furniture rather than a machine, so the radio was accepted into more homes once it began to change its appearance and was able to be bought off the shelf. The development of valve radios and loudspeakers improved the quality of reception, while the spread of reliable electrical power obviated the need for bulky and unreliable batteries. Radio prices fell further, and small, cheap units came onto the market. The sonic quality of broadcasting was further refined by the advent of electronic recording and playback technology in 1925.[11] These technological developments allowed radio listening to become a communal experience, as did the local manufacture of sets, which cut their price. By the late 1920s, despite the onset of the Depression, radio was the 'electronic hearth' and an axis of family life in homes around New Zealand.

It was also another exciting and mysterious technology of acoustic modernity that brought new sounds into people's lives in different ways from records and films. Gramophone records provided listeners with the same sounds but heard at different places and times. The talkies provided the same sounds experienced simultaneously but only in one place. In this way, records and talkies resembled newspapers, but with smaller audiences. Radio listeners did not constitute a crowd in a physical space such as a theatre, but they did form an audience.

Benedict Anderson has pointed out the role of newspapers in making nations or 'imagined communities'.[12] Radio came to play an even more important role than print media in this sense of national communion, by bonding its listeners together through the simultaneous experience of acoustic events. These could be anything from the live coverage of a wedding to a relayed speech by a newly crowned king on the other side of the world.[13] The media theorist Rudolf Arnheim made this point in the context of a discussion on radio and national politics in 1935, when he wrote that 'wireless without prejudice serves everything that implies dissemination and community of feeling and works against separateness and isolation'.[14] The mass movements of the 1930s such as fascism and communism were preoccupied with, among other things, removing 'separateness and isolation' within nation states. The simultaneously uniform experiences of radio sounds for radio listeners were important components of this imagined unity.[15] But this does not mean that these uniform sounds were heard in a uniform way.

Imagination was (and still is) a key element in radio listening. Radio historian Susan Douglas points to the agency of listeners in making 'imagined communities more tangible', as well as the important role of imagination.[16] She outlines three radio listening styles — informational, dimensional and associative — that people developed as the medium became a regular part of daily life at home and in public.[17] Informational listening involves taking in facts but using little or no imagination, such as when hearing sport results or weather forecasts. Dimensional listening invokes the cliché of radio as the theatre of mind. Radio serials, Clive Drummond's report of Aunt Gwen's wedding and the radio commentary of a rugby match are all broadcasts that evoke images and emotions through the imagination. Douglas calls the third style of radio listening associative, since it builds on a listener's own network of links and networks associated with sounds. A song or voice heard on a radio can often trigger other memories associated with it.[18] An individual's radio listening might be constantly shuffling among these styles at any given time. And they all involve intentionality in that they are directed at sounds in an attentive manner.

Tensions between engaged and passive listening were part of New Zealand's radio culture between the wars, when radio first became a part of most people's daily lives. Some saw an intentional and attentive mode of listening as the ideal way of hearing radio, in that it allowed for the possibility of a new style of cultural expression to be developed. Treating radio sounds as background noise could have dire consequences, according to Arnheim, who saw such listening as a deadening to an individual's freedom and inner life. In his view, radio in the home was a danger because it reduced artworks to soothing sounds in the background of domestic life. To forestall this, radio listeners needed education and self-discipline so that they listened appropriately to the artworks that the medium lifted out of the theatres and concert halls and placed in the living room.[19]

The styles of listening discussed by Douglas and Arnheim have in common the view of radio as a medium that stimulates a listener's imagination. The lack of images frees a listener's mind to create their own responses, associations and mental images. The radio listener becomes the producer, director, designer and stage manager of the imagined spectacles and situations that radio sounds produce.

•••

In New Zealand, as elsewhere, radio was mostly heard in the home between the wars. By the mid-1920s, the receiver's machinery was housed in attractive cases designed to complement the style of the domestic interiors of the time.[20] These cases were often built from new materials like Bakelite and in the modernist style now known as Art Deco.[21] Radios might bring the style as well as the sounds of the modern world into the home.

But many radio sets also used traditional styles and materials which added to the medium's respectability. In a photograph of a New Zealand dining-room from the 1920s, the radio sits on a cabinet to the left of the fireplace. The receiver's case is built from a wood that matches the other wooden furniture in the room. The speaker on top of the case is of modest size and does not inconveniently protrude into the room, as did early phonograph horns. There are no visible wires or batteries to detract from the radio's discreet appearance. Apart from the speaker, it might be taken for a portable writing desk. A comfortable chair sits next to the radio for the convenience of a listener.

In another photograph, a living-room interior from 1938, the radio is next to the fireplace, highlighting its role as the original 'electronic hearth' before the rise of television. The design of the radio is sleeker and more streamlined than the one in the earlier photograph. The controls on the

This interior of a dining room from the 1920s shows the radio positioned on a table next to the fireplace. The cabinet design and finish match the other furniture in the room. Alexander Turnbull Library, Wellington, Ref: 1/1-010493-G

radio are also easier to use. There is a clear tuning-band display that allows for easy station selection by the young man. The older man and the woman in the photo are arranged in an archetypal radio-listening pose. They are apparently at ease and seem to be engaged in reading and sewing while still registering the radio's sounds. In the mirror can be seen another man who is standing and looking at the three people in the frame. He may well be the director of this scene that seems carefully curated to illustrate one idea of what listening to the radio meant for many people between the wars.

These images may or may not be staged, but other sources suggest that many New Zealanders placed radios in the centre of domestic life. The author Barbara Anderson recalled her family's 1938 purchase of a large Gulbransen radio set that was almost as tall as she was. The radio 'stood in the sitting room between the padded window seats and Grandfather Wright's bookcase'.[22] Lorraine Russell was an Auckland schoolgirl in the late 1930s when her family bought a radio that 'sat in pride of place in the middle of the mantelpiece'.[23] This may have provided new entertainments, but Lorraine wrote that the arrival of the radio ended 'a part of my life I enjoyed so much. I would no longer sit on Dad's knee in the evenings and sing the old songs'.[24]

Radio as the home's electronic hearth, 1930s. Unlike television, radio does not dictate the arrangement of chairs in a room. Alexander Turnbull Library, Wellington, Ref: 1/1-017566-F

The structures and habits of domestic and family life began to be shaped around the schedules of the radio, and not all of these were welcome.

A 1936 letter to the *New Zealand Woman's Weekly* signed by 'Gwen' accused radio broadcasting of dividing families, ending singing and playing music in the home, and depriving children of the personal touch of parental storytelling. For Gwen, radio in the home was an anti-human technology that dulled people's lives and destroyed the place where people 'could retire and quietly contemplate, could rest calmly'. Beneficial activities such as learning to play an instrument or reading to a child were abandoned in favour of canned offerings. The walls of the domestic citadel had been breached by the radio 'that vomits forth its everlasting mixture of soundwaves'.[25]

'Hairini', however, argued that listeners seldom tuned in all day, so the radio was hardly 'everlasting'. Hairini went on to add that radio brought entertainment and information to people living in remote country areas and that the music on the radio was an incentive for young musicians rather than a substitute. For this writer, home life was enhanced by the radio. Hairini also listed features appreciated by 'the rational listener', such as news, music, sport, church services, Empire broadcasts and 'last but not least, the wrestling' – a sly reference that undercut the pretensions of Gwen's letter.[26]

Negative views of radio seem in fact to have had little support. By 1936, when these letters were written, 192,600 New Zealanders had obtained radio licences. This number increased by 50,000 in 1937.[27] All around New Zealand, domestic activities were being built around the structures and rhythms of the radio schedules.

•••

Early radio broadcasts were unpredictable and sporadic, but gradually became more organised and predictable. This process accelerated after the formation of the RBC, which coordinated nationwide transmission.[28] Listeners needed to know what was on, when it was on, and what frequency it was on. Newspapers began publishing radio schedules in 1926.[29] These were the ones most referred to on a daily basis, but the RBC understood the need to give people more than the sort of basic information found in the daily papers. On 22 July 1927 the RBC began publishing a weekly magazine, the *N.Z. Radio Record*. This was the direct ancestor of the *Listener* and was modelled on the BBC's *Radio Times*, which began in 1923.[30] The *Radio Record* featured articles about broadcasters, radio stations, radios, music, and general information about the RBC and its doings. It also carried full weekly schedules for each of the four YA stations. Some B stations published their schedules in newspapers, but they had difficulty in publicising their programmes due to lack of funds. With the *Radio Record* and the stronger signals of its own stations, the RBC dominated broadcasting in New Zealand throughout the late 1920s.

The publishing of regular programme schedules was critical to the social invention of radio listening. The broadcaster's expectation was that people would select which items they wanted to hear and tune in just for those. The conditions and difficulties of listening were such that broadcasters did not expect listeners simply to leave the radio playing at all times. The RBC programmers endeavoured to coordinate their broadcast content across their four YA stations so that, 'as far as possible, no two stations are broadcasting the same type of programme the same evening'.[31] Initially they had to fill roughly seven hours of airtime each night, with broadcasts usually beginning at about 3pm and ending at 10pm. These might go for longer over the weekends. Each YA station also had regular silent days during which maintenance was carried out on their equipment. Broadcasting hours were expanded during the 1930s, but even by 1940 radio stations shut down overnight and might still only broadcast for several hours in a day.

Given the idea of public service, the RBC broadcasters aimed at variety from the start. For example, on Tuesday 13 September 1927 Wellington's 2YA began broadcasting at 3pm with a gramophone recital.[32] A lecturette on 'Gas Cooking' by Marion Christian followed at 3.30 until 3.45, when more gramophone items continued until news and market reports from 7pm until 7.30pm. N.R. Jacobson lectured on 'Air and its Wonders' before the 8pm chimes of the General Post Office clock heralded an hour of live musical items, including instrumental solos and songs. There was a one-minute weather report at 9pm followed by more live musical items, among them a relay of an orchestra performing at a theatre. Douglas Tayler gave an illustrated lecturette on 'Music and Pictures' for 10 minutes at 9.25pm – the illustrations were presumably musical. More live musical items continued until the station shut down at about 10.20pm.

On the same day, 3YA in Christchurch was silent. 1YA in Auckland had a children's session with Aunt Betty from 6.30 to 7.15pm; a talk by Stanley Bull on 'Yosemite Valley and Grand Canyon'; a relay of the Majestic Theatre Orchestra and two comedy sketches by the Griffiths Duo, along with instrumental and vocal items performed live in the studio until the 10pm 'Close Down'. Dunedin's 4YA began at 3pm with a gramophone recital until, at 3.15, M. Puechgud continued her series of lectures on interior design with one about 'Homes from the Outside'. Studio music filled the half hour from 3.30 until 4pm, when H. Greenwood, the librarian of the Dunedin Athenaeum, reviewed a selection of books. The station closed down from 4.30 until 7pm, when a half-hour children's session took listeners to the news at 7.30. At 7.45 Pastor W.D. More gave a humorous address appositely entitled 'The Unlucky Thirteenth'. A relay from His Majesty's Theatre of light classics filled the two hours from 8 until shutdown at 10pm.

This was a typical day of radio broadcasting for the RBC and New Zealand listeners. The main emphasis was on the late afternoon and evening hours. Radio was listened to after work or school, so schedules were designed to produce a mixture of information, education and entertainment during those times. The rhythms of radio broadcasting both synchronised with and affected the rhythms of domestic life.

George Clark recalled the great popularity of radio serials when he was a boy in Ngatea during the 1930s. Not everyone had a radio in their home and so 'people would arrange to be at a certain place to hear their favourite radio serial'.[33] Dawn Sheppard was a young girl living in Otago during the late 1930s and she too used to wait impatiently for the 2YA children's

This advertisement for the debut episode of the radio serial 'Fred and Maggie Everybody' twice mentions the 7pm timeslot, highlighting the urgency of being in the right place at the right time to hear the new radio serial.
Auckland Star, *4 October 1937, 16*

session to begin. When she was a little older, she would dash home from school at lunchtime so her mother could update her on the morning's serials. The Australian-made serial 'Fred and Maggie Everybody' was on every Monday and Tuesday at 7pm, and Dawn seldom missed it.[34] Station 1ZB made sure listeners knew the 'Fred and Maggie Everybody' timeslot by placing a series of cartoons in newspapers on the first day the serial went to air. This serial became immensely popular and over 40,000 people turned out to see the show's stars who visited the Wellington Centennial Exhibition in 1940.[35] Radio stars were major celebrities.

Meal times, bed times, bath times and other regular moments in people's private lives were disrupted and rearranged by the regular scheduling of radio broadcasts. Radio extended public time into the household in ways that no other technology had before it.[36] Household routines had been ordered around publicly shared time settings so that workers could be at their jobs and children at schools during various parts of the day. But outside these hours, domestic time was ordered according to the desires and needs of the household members. Radio, however, placed home leisure time in the same sort of timetabled system that organised work hours and transportation. Appointment radio listening was similar to catching a train or getting to work on time. A person's job or commuter train went on regardless of whether he or she was there, just as 'Fred and Maggie Everybody' was broadcast whether Dawn Sheppard heard it or not. Programming organised regular listening habits and made them predictable. It also raised questions about what was scheduled and why it was there.

Radio scholar Paddy Scannell has described the BBC's programming in the early 1930s as 'designed to encourage attentive listening and to discourage the lazy listener. The idea then was to cater for all tastes, minority and majority, over a period of time – a month was the period sometimes quoted.'[37] This same idea was applied in New Zealand, but the focus was not so much on teaching people how to listen or discouraging 'lazy listeners', but rather to give listeners a fair return on their 30-shilling annual licence fee by providing them with a range of entertainment that would cover most tastes over the course of a week. The BBC's ethos was based on the principles famously formulated by John Reith, the organisation's first general manager, as 'Inform, educate, and entertain'. The idea was to shape society by promoting culture, civilisation and democracy.[38] These principles were also part of the ethos of the RBC, but public service rather than public improvement was a more important component of its policy. As the RBC's manager A.R. Harris stated in 1928:

> That policy is to provide the highest quality of entertainment obtainable from the talent available in New Zealand and disseminate it in such a way as to bring it within the range of modern receiving sets in any part of the Dominion for a nominal annual licence fee, and to broadcast a service based on a spirit of co-operative effort designed to promote the best interests of all sections of the community.[39]

What constituted the 'highest quality of entertainment' was the topic of an ongoing debate between the RBC and its listeners, as well as among the listeners themselves. It generated many discussions in the *Radio Record*'s letter pages. One such contribution, by W.J.K from Dunedin, described the sort of listening that the RBC programmes were designed for:

> I think the wireless gormandizer is a principal [*sic*] cause of the discussion one hears about the programmes. Night after night he listens in to a radio programme. Can any human mind stand such surfeit? Or to have to listen to twelve or fourteen hours of the best canned music on the gramophone

throughout the week would, I think, not be tolerated by anyone. Personally I'd be kicking the instrument into the backyard about halfway through the second evening. Why not take it in moderation and listen when the mood and the opportunity is right?[40]

Too much radio was not a good thing, and the RBC and its subsequent organisations tried to offer variety so that listeners would not get bored and jaded. The reference to 'canned music', meaning recorded music, is significant. Music was at the centre of many of the discussion, debates and outright arguments about what was to be heard from New Zealand's radios.

•••

The musical items broadcast on New Zealand radio between the wars were a worthy and uplifting mixture of light classics, religious music, truncated and arranged classical pieces, brass band music, whistling and songs. They reflected a repertoire that was favoured in middle-class homes with pretensions to respectability. The music made a few concessions to the popular taste of the time with its dance crazes, sentimental ballads, show tunes and jazz, but it also avoided the esoteric plateaux of rarefied high art. 'Middlebrow' would have been the contemporary term to describe it.

Radio listeners heard far more live music than recorded music during the 1920s. One reason for this was that the 1925 Radio Regulations were interpreted to mean that recordings could comprise a maximum of 25 per cent of transmission time. Several reasons for this have been suggested. Broadcaster Ken Collins believed that it 'was to ensure that local talent was properly fostered'.[41] Given the state of radio in 1925 and its minor role in public life, this is highly unlikely although may have happened to some extent. It is more likely that the recording industry, fearful of radio and its effect on sales, lobbied to have this clause in the regulations. Also, electric pickups became available in New Zealand only in 1928 and were in widespread use only by 1930. Up until then, recordings were broadcast by the simple expedient of holding a microphone next to the gramophone, and this resulted in very poor quality sound.[42]

Some of these musicians performed for the pleasure of being on the radio but the subject of remuneration became important and was related to the question of professionalism. It was felt that a regular and appropriate system of payment would cut out the amateur talent and raise standards of performance. The question was sharpened by the increasing numbers of musicians made unemployed by the talkies. By 1930 a standard system of contracts and payments was in place, and that year the RBC paid out £16,000 to performers.[43] Finding enough competent musicians to supply this music as well as the money to pay them was a problem for the RBC. The problem for listeners was that there was a lack of variety in the broadcasts and the quality of the performers varied. As 'A to Z', writing from Marlborough pointed out: 'Our local artists are quite good, some quite first class, a good many quite second rate, but I think all must agree the artists to be heard from gramophone records are nearly all 100 per cent excellent.'[44]

Most people who expressed opinions on this subject through the pages of the *Radio Record* or newspapers tended to agree that it was impossible for radio to please everyone all of the time but still wanted the broadcasters to play more (or less) of whatever music the writer felt strongly about. As expressed, this debate tended to be polarised between 'lowbrow' and 'highbrow' or 'jazz' – which meant popular music in general – versus 'classical'. The following are two typical examples from 1928:

> In reference to the programmes put on by the New Zealand Broadcasting Company, I must and can only say that they are getting worse, and the cry for cutting out the 'highbrow stuff' is left unheeded. The general public to my knowledge do not want highly classical music night after night, and while I don't want it myself, I can put up with it for a couple of nights a week provided light stuff or popular music is given more often. The Broadcasting Company expect more listeners. How can they get them unless they provide entertainment for them?[45]

> Would it not be possible to put on some music instead of jazz rubbish? A jazz record now and again can be tolerated, but it is awful to wait half an hour and more listening to the thump, thump of a jazz band, accompanied by a singer (?) with a voice like a starved tom-cat. Surely it would be possible to mix the items a little. Hope you take this as a friendly suggestion. I've growled more than I intended because I'm feeling blue after all that jazz.[46]

The debates about jazz and classical music were indicative of the problems faced by New Zealand broadcasters between the wars as they developed a radio service that was national in coverage and presented the 'highest quality in entertainment' to those who had paid the not insubstantial sum for their licence. These debates were enacted at the administrative level of the RBC itself with the appointment of J. Bellingham as director of music in September 1927. He saw the role of radio in Reithian terms:

> I can conceive of no greater medium for the intellectual and artistic uplift of the nation than broadcasting. While all good music must have a refining and elevating influence, it cannot be overlooked that certain music is apt to have the reverse effect. As an example of the harm that may be done to the community by music, I would point to the hold that jazz has had on the American people. It is our aim to entertain and educate. That is to say, while we will not lose sight of the entertainment factor, we will endeavour to eliminate that which is harmful.[47]

Bellingham quickly set to work on this project of cultural inoculation against the musical disease of jazz with the formation of permanent classical groups attached to 2YA (the RBC's most powerful station) and the promotion of classical and opera broadcasts. Information and publicity about the new programme policy heavily featured in the *Radio Record* as part of his strategy to enlighten as well as entertain listeners.[48]

However, Bellingham's stance was in direct opposition to that of the RBC's manager, Ambrose Harris. Harris was interested in making programmes that would draw more listeners and therefore stimulate the sales of radio sets and increase the RBC's income from licence fees. Harris did not see the RBC as an institution that should work along Arnoldian lines to 'raise' the quality of the nation's musical life by presenting people with established classics that were meant to be good for them. It was Harris who triumphed. Bellingham resigned and went to England in June 1929.[49] This did not mean that the RBC suddenly played nothing but jazz. Harris may have wanted to broadcast more popular music, but the political and social elites who governed the RBC ensured that music programming remained the same through the RBC's existence and persisted when it was replaced by the Radio Broadcasting Board in 1931, which was in turn superseded by the New Zealand Broadcasting Service in 1936. Radio in New Zealand was to cling to its role of cultural uplift for many more years, and this ideology still persists to some degree in the aims and activities of the present-day public broadcaster, Radio New Zealand, although entertainment is more important and the cultural uplift consists of cultural inclusiveness.

• • •

The sonic and temporal organisation of radio listening through programming and schedules offered a range of sounds to New Zealanders in the inter-war years, and they could pick and choose as they saw fit. But it was not always the case that the sounds heard were the ones in the schedules.

Radio listening sometimes involved an effort to sort out what was signal and what was noise. The cartoonist was clearly sceptical about the new technology as an educational panacea. Auckland Star, 8 September 1928, Supplement, 3

Unwanted noise in the form of hisses, buzzes, hums or static plagued New Zealand's radio listeners. In 1933, the Post and Telegraph Department surveyed the 1249 complaints about radio interference it had received in six months.[50] The causes included trams (54), street lighting (67), trees touching power lines (22), amateur radio transmitters (35) and so on. The most common culprits were power lines and fittings (214). With so many potential causes of interference, static became part and parcel of radio listening, just as early gramophone listening had meant taking surface noises on board. Betty Mules was a child in Ngatea during the 1930s and remembered that 'the reception was always crackly'.[51] Patricia Ridding, growing up in Wellington, heard the family radio produce 'many crackles'.[52] A cartoon from the period draws attention to the amount of static involved in radio listening while also winking at the idea of radio as a serious medium, or what a 1927 *Radio Record* editorial called a 'machine that can be turned to greater good'.[53]

Apart from the annoyance of static, some people heard radio as an irritating noise in itself, at least when it was too loud. Noise abatement movements, or campaigns to turn down the noise of urban living, were common during the twentieth century, especially in the large cities of America, and often singled out radios.[54] The speakers used with radios from the mid-1920s onward were the real cause of the problem. Thus, in New Zealand too, some people were ready to hear the radio as noise pollution. D.G.C. wrote in the *New Zealand Woman's Weekly* in 1936 about the unhealthy effects of excessive noise such as disturbed sleep, increased fatigue, impaired hearing, decreased efficiency at work and interference with the development of children.[55] The article cites the neighbour's radio as part of a typically too-noisy day. The writer also maintains that 'radios should be kept low in both shop and home' to avoid music and announcements 'booming along a suburban street'.

The dangers of excessive noise and the role played by too-loud radio were the subject of numerous letters and articles between the wars.[56] The subject was even raised in a 1935 debate on the Broadcasting Amendment Bill in the Legislative Council. The Wellington representative, George Andrews, moved an amendment that would penalise anyone found guilty of causing nuisance with 'a wireless instrument'; the guilty party would be liable to a fine of not less than £2 and no more than £5, with at least £5 for every subsequent offence. Andrews claimed that some people disturbed their neighbours by playing radios early in the morning and late at night with 'mainly jazz tunes – one certainly cannot call them music – and they are enough to drive one silly'. The problem was particularly bad in small towns and where young people took holiday baches and 'kept the radio running until the last Australian station shut down'.[57]

As it happened, Andrews' amendment was not accepted, as such matters could be dealt with under the Police Offences Act. But Andrews was not just attacking loud radios; he was also objecting to what they played and who was listening. He, like many older people throughout the twentieth century, found that the intersection of young people, popular music and audio technologies was a potential site of instability and rebellion.

The 'how' of radio listening was not immediately obvious when the medium arrived in New Zealanders' homes from the mid-1920s onwards. The hardware was rapidly domesticated to fit snugly and attractively into domestic spaces. Programming schedules established reliable timetables and led to regular listening habits which helped people get to grips with the medium. It soon became established as an informative, entertaining and very enjoyable pastime that was a regular part of the daily round. But what was on the radio and what did it mean to listeners?

Chapter 9

ORGANISING RADIO

•••

Deciding what sounds should be heard from radios interested both broadcasters and listeners. The most important attempt at finding out what listeners wanted to hear was the plebiscite undertaken in 1932 by the newly instigated New Zealand Broadcasting Board (NZBB), the successor to the RBC. Between March and the end of May, the NZBB distributed 50,000 questionnaires to radio licence holders, of which 24,000 were completed and returned.[1] Given that there were 75,000 radio license holders in New Zealand in 1932, the survey was limited from its inception.[2]

It may have been imperfect in many ways, but the survey was the first of its kind in New Zealand, and the data from it is valuable as a sounding of what people wanted from radio as it became a normal part of everyday life. It is also important in that it provides a rough indication of popular music taste in New Zealand at the time. There is very little reliable historical data about this: there were no record charts, and sales figures for 78s and sheet music are very rare for this period in New Zealand. The 1932 survey gives us some idea of what people wanted to hear.

The survey's six questions were mostly multi-choice with no room given for suggestions or opinions. The questions were about recordings versus live performances, the sort of wireless being used, which station was most listened to, the preferred hours of listening, the type of preferred programme or sessions, and the preferred form of entertainment. But the possible answers were supplied by the NZBB, which caused some criticism from both listeners and editorial writers, who claimed the survey was a set-up and something of a stunt on the part of the NZBB.[3]

The NZBB publicised the survey through the press and the *Radio Record*, and stated that it was intended to help the board 'balance its programmes in accordance with the wishes of its licensees'.[4] The idea was to hear from the so-called 'silent listener-in who does not bother to register either appreciation or condemnation of the present broadcast entertainment', and that 'no querulous minority will be unduly favoured'. So the survey forms went out. The board gave listeners from March until the end of April to fill them in, but extended the time until the end of May to ensure the largest possible number of replies. Along with the questionnaire forms, many people enclosed letters in which they fleshed over the bare bones of the plebiscite's options. One folder of these survives in Archives New Zealand.[5] They indicate that radio listening was becoming a mainstream activity by 1932. As listener H.N. Bullock wrote, a wireless set had 'ceased to be a freak or a play-thing'.[6]

After some time spent tabulating and adding the results from the 24,000 forms returned, the NZBB published its research. In summary, this showed a preference for recordings over live performances by local artists; most people used valve radios, not crystal sets; 2YA was the most listened to station; 8pm to 9pm was the hour most frequently used for listening; evening concert sessions were the most popular; and band music was the preferred option for entertainment.[7]

It should be remembered that radio broadcasting then was not the 24/7 ceaseless and continuous wall of sound we are used to: in modern radio, 'dead air' (silence) is anathema. But in 1932 stations might broadcast for only certain parts of the day and only on certain days of the week. Rest days were rostered across the NZBB-owned stations to allow maintenance. Reception could be patchy and sometimes a matter of luck. Announcers spoke formally, clearly, slowly and with English accents, and there were clear pauses between each element of the broadcast, be it music or a serial or a news bulletin. It is important not to project our modern idea of radio back onto the past. The sound of radio past was radically different from the sound of radio present.

The sort of music heard from the radio was constantly negotiated between radio makers and radio listeners. During the 1920s and into the mid-1930s, live performances from local artists formed the majority of the musical content of New Zealand's radio broadcasts. The 1932 survey asked listeners whether they preferred recordings or local artists, and recordings were the clear winner, with 74.11 per cent of the votes.[8] A typical reason for this preference was that records offered 'the pick of the world's artists'.[9] 'Listener in' of Wellington liked hearing records, as few of the world's top

RADIO POLICY.

BOARD SEEKS ADVICE.

PLEBISCITE OF LISTENERS.

To ascertain the views of listeners regarding programmes and other matters of interest, the Radio Board has decided to take a plebiscite, and the forms on which the questions are printed are now obtainable at the Post Office. When filled in the forms are to be sent to "The General Manager, New Zealand Broadcasting Board, Wellington," who will pay the postage.

The plebiscite is understood to be one of the first ever taken by any authority controlling broadcasting to ascertain the wishes of listeners. Fifty thousand forms have been sent out.

Following is the text of the form:—

Kindly aid the New Zealand Broadcasting Board and the service by completing this form.

State your Location.
(Nearest Post Office.)
....
Do you prefer entertainment provided by—
  1. Local Artists
       or
  2. Recordings.
(Strike out the item for which you do not intend to vote.)
For reception, do you use—
  1. Crystal
       or
  2. Valves. (How many?)........
(Strike out the type NOT used.)
To which New Zealand Station do you listen most frequently?...............
If the general broadcasting hours were 10 a.m. to 11 p.m., at what times would your set normally be in use. (Indicate by means of crosses underneath the periods set out below.)

10 to 11 a.m.	11 to 12	12 to 1 p.m.	1 to 2 p.m.	2 to 3 p.m.
3 to 4 p.m.	4 to 5 p.m.	5 to 6 p.m.	6 to 7 p.m.	7 to 8 p.m.
8 to 9 p.m.	9 to 10 p.m.	10 to 11 p.m.		

cont'd next column

By means of numbers on the lines on the left indicate the order of your preference for the following sessions. (Your first preference should be numbered 1). Indicate whether you consider the length of the respective sessions should be increased, decreased, or remain unaltered, by striking out two of the terms (more, less, satisfied) on the right.

.... Religious services more less satisfied
.... Children's session more less satisfied
.... News and market reports more less satisfied
.... Evening concert session more less satisfied
.... Running descriptions of sporting events more less satisfied
.... Dance session more less satisfied
.... Dinner session more less satisfied

Indicate by means of numbers the order of your preference for the following classes of entertainment during the evenings:—

.... Hawaiian music.
.... Elocutionary items (serious and dramatic).
.... Elocutionary items (light and humorous).
.... Short plays and sketches.
.... Talks (educative and informative).
.... Band music.
.... Light orchestral items.
.... Classical and chamber music.
.... Opera and oratorio.
.... Whole-evening plays.
.... Humorous songs and monologues.
.... Comic opera and musical comedy items (vocal).
.... Concerts by musical societies and choirs.
.... Vocal solos, duets trios, and quartets.
.... Talks (sporting and topical).
.... Instrumental solos, duets, trios, and quartets.
.... Community singing.

The 1932 Plebiscite of Listeners questions were designed to find out what the public wanted to hear – within limits. Press, 16 March 1932, 13

artists came to New Zealand which 'is far away from the musical centres of the world'. The high price of 'good records' also meant that he or she 'looked to the radio to give us what we cannot get otherwise, the very best recorded music'.[10] For 'Listener in', the radio was both an ear on the world of music beyond New Zealand and a way of gathering information about 'good' and 'bad music'.

Most of the survey's respondents were clear about when they wanted to listen in and hear music. They wanted music from 6pm until 10pm, with a dinner session running between 6pm and 7pm. Many letter writers preferred light music for the dinner session. D. Murray of Tuhua, near Taumarunui, wanted 'less high class music and less jazz and more music that the ordinary citizen can appreciate'.[11] The music preferred for dinner hour did not make demands on its listeners' attention. As one respondent put it, 'the dinner hour music particularly might be to the cultured ear the finest, but to me it takes a lot of following and seems to give my set indigestion'.[12] These listeners were not against 'high class music' in itself but rather were suggesting that it was not appropriate at a time when many people might be eating.

The development of programming such as the dinner sessions was part of the process that entwined domestic time around radio schedules. The most preferred listening hours of 6–7, 8–9 and 9–10pm also indicate the role of radio as the original 'electronic hearth'.[13] Listening was done in living rooms, and often with neighbours or friends as well. The lack of enthusiasm for listening during the morning hours made midday the obvious time to begin transmissions.

The responses to the survey show how alien radio in 1932 would seem to modern ears. The limited transmissions hours, patchy reception, and radio's role as the centre of household entertainment are all important aspects of historicising the medium. The past practices, conventions and concepts referenced by the word 'radio' are radically different from modern ones. The listening experience itself was quite different and needs to be understood on its own terms as part of the fabric of lived daily experience. Silence itself was an inherent part of radio between the wars, and is a stark contrast with the modern practice of 'always on, always there, always sounding' radio styles. The 1932 survey and the commentary it generated highlighted these differences, and prompted discussions about what was appropriate for radio and the wider culture in general. Several letters written to the *Evening Post* about the survey clearly show how different listeners heard radio in different ways.

Summary.

Question 1: "Do you prefer entertainment provided by—
Local artists
or
Recordings?"

Result:
	Per cent. of votes.
Local Artists	25.89
Recordings	74.11

Question 2: "For reception, Do you use Crystal
or
Valves?"

Result:
	Per cent.
Crystal	6.42
Valves	93.58

Question 3: "To which New Zealand Station do you listen most frequently?"

Result:
	Per cent.
1YA	16.06
2YA	57.96
3YA	9.92
4YA	5.35
Others	10.71

Question 4: "If the general broadcasting hours were 10 a.m. to 11 p.m., at what times would your set normally be in use?"

Result (in order of preference):
1st .. 8 p.m. to 9 p.m.
2nd .. 9 p.m. to 10 p.m.
3rd .. 6 p.m. to 7 p.m.
4th .. 7 p.m. to 8 p.m.
5th .. Noon to 1 p.m.
6th .. 10 p.m. to 11 p.m.
7th .. 3 p.m. to 4 p.m.
8th .. 5 p.m. to 6 p.m.
9th .. 4 p.m. to 5 p.m.
10th .. 1 p.m. to 2 p.m.
11th .. 2 p.m. to 3 p.m.
12th .. 10 a.m. to 11 a.m.
13th .. 11 a.m. to Noon

cont'd next column

Question 5: "Indicate the order of preference for the following sessions."

Result:
1st—Evening Concert Session.
2nd—Dinner Session.
3rd—Description of Sporting Events.
4th—Dance Session.
5th—News and Market Reports.
6th—Children's Session.
7th—Religious Services.

Question 6: "Indicate the order of your preference for the following classes of entertainment during the evening."

Result:
1st—Band Music.
2nd—Light orchestral items.
3rd—Humorous Songs and monologues.
4th—Comic opera and musical comedy items (vocal).
5th—Instrumental solos, duets, trios, and quartets.
6th—Hawaiian music.
7th—Community singing.
8th—Vocal solos, duets, trios, and quartets.
9th—Talks (educative and informative).
10th—Elocutionary items (light and humorous).
11th—Short plays and sketches.
12th—Concerts by musical societies and choirs.
13th—Talks (sporting and topical).
14th—Opera and Oratorio.
15th—Classical and Chamber Music.
16th—Elocutionary items (serious and dramatic).
17th—Whole evening plays.

The people talked back to the radio in the form of the plebiscite results. Band music was in and long, heavy drama was out. Press, 7 September 1932, 5

When the survey was announced, one listener who used the pen name 'Simeon Stylite' was quick to deliver some fairly ascetic views to the editor of the *Evening Post*. Simeon wrote that asking people what music they wanted on the radio was the same as asking children what they should be taught. People needed guidance in their choices just as children need teachers. In Simeon's view, the survey represented a dangerous threat to the nation as 'a high standard of musical taste is essential to the wellbeing of a civilised community and no artistic progress can ever result from giving the public what they want'.[14]

A response from 'Ballot Box', a writer of a more democratic bent, was printed the next day. Ballot Box suggested that Simeon should appoint a cabinet of his own choice to run the country in all matters, given he had so little faith in majority rule. Ballot Box also looked forward to the survey results showing what music people wanted and 'the sad, sad, ending to the orgy of highbrow music he [Simeon] has no doubt enjoyed in the past'.[15] Sarcastically declining Simeon's 'kind offer to improve my musical taste', Ballot Box went on to ask why 'lovers of modern music have every night to wait until 10pm to hear that portion of the programme', and hoped that Simeon might grow more fair minded as he or she got older and realised that those who pay the same fee are entitled to the same treatment.

Simeon was not going to let it go at that, and replied to Ballot Box the next day. Simeon remarked on Ballot Box's preference for dance music, and wrote quite waspishly: 'Music of one syllable may be very well for the unsophisticated, but one would be sorry to learn that a majority of music lovers in this country desired nothing better to listen to than such a poor substitute for the real thing.'[16]

This was obviously an enjoyable affray for all concerned and reminds us that trolling was alive and well long before the internet. But the exchange illustrates the cultural politics, the lines that were being drawn as radio became part of the fabric of everyday life. What was radio for, exactly? What do we do with it? What are appropriate sounds to introduce into the sanctified privacy of the home? The division between home life and work life was much stronger and wider in those days than it is today. The question of what influences might enter home life through the latest piece of handsome domestic furniture, the wireless, was a very serious one.

Naturally everyone was thinking of the children, and the fears that had arisen from the advent of the talkies were amplified by the radio. Children were learning slang, bad words and terrible enunciation from the talkies, but at least the youngsters could be forbidden to go to the cinema. But how could a concerned parent cope if deep within the safety of the domestic bosom lay the beguiling viper of the radio? Placing the radio in a public room was one solution, but a better one was to make sure that vulgarity and harmful moral pollutants such as slang and jazz were not on the radio in the first place.

• • •

Fears about the harmful effects of media on society, family life and young people in particular are still with us in various forms. But the young people and children of the 1930s were avid consumers of the new media, and entered the exciting new soundscapes opened up by radio with enthusiasm.

For the very young, radio was a source of fun and delight. Sessions for children began as transmission hours increased, and were usually placed at 6pm to 7pm timeslots. The format for these sessions included stories, songs and jokes (often sent in by and performed by children), with adults supplying the announcing.[17] The adults on children's radio were called 'uncles' and 'aunts', like Aunt Gwen on the 2YA session. This gave them a form of authority that seemingly placed them within a listener's family structure, and was warm and not too threatening.[18] Uncles and aunts promised mischief and fun.

The programmes made by these presenters were often quite different from those made for adults. Unpredictability, inventiveness and humour were deployed to keep children interested. Silly voices and stories were very popular and made children's radio truly innovative. Historian Lesley Johnson, writing of the Australian experience, claims that 'the emphasis on spontaneity, on the unrehearsed, on the relaxed and the friendly, and the invitation to young listeners to enter this imaginary world, exploited wireless broadcasting as a new medium more fully than did other early programmes.'[19] The same could be written of New Zealand.

Much of the format and content of New Zealand's children's sessions was copied from the BBC's 'Children's Hour', which first went to air during 1923. The 'aunts' and 'uncles' nomenclature was first used on 'Children's Hour', as were other innovations including the first stories read over the radio, the first play written for radio and the first use of birthday greetings.[20] These were introduced to New Zealand by Ernest Bell, head librarian at Christchurch Public Library.[21] As 'Uncle Jack', Bell was heard twice a week for over a decade on Christchurch station 3YA's children's sessions. His co-host, Aunt Edna, was a library assistant named Edna Pearce.[22]

Bell estimated the session received over 8000 messages, letters and telegrams in just two years from 1926 until 1928.[23] Many of these were birthday requests, or 'shoutouts' in the modern phrase. Recognising the excitement for children of hearing their names on the radio, other stations followed Bell's lead and invited listeners to write in. Fred Price of Auckland was one who sent a birthday greeting request to Auckland's 1YA on 3 May 1932. Price wrote that Shirley, his daughter, was turning six on 5 May, and that 'we propose to put a present in the wireless cabinet and also under her pillow'.[24]

As a boy living in Otago during the 1930s, a highlight of Donald Webster's day was the 4YA 'Children's Hour' run by Pastor William More under the name of 'Big Brother Bill'.[25] Webster remembered his enjoyment of the songs and stories and particularly a regular section about model aeroplanes. He also recalled that birthday greetings were read out at the end of the session and might include instructions such as: 'Bruce, if you follow the string tied to the wash-house door you will find your surprise in the garden.'[26]

A similar sense of excitement was caught in an email one listener sent to Jim Sullivan's 'Sounds Historical' on Radio New Zealand in 2008. Mary Matthews did not give the name of the town she lived in as a child but she had very strong memories of radio during the 1920s. One day, while sitting on her father's knee, she heard her name and a 'happy birthday' coming through the speaker. The voice added that 'the radio fairies had left a present … on the battery box'. She was very surprised. But Mary and her sister were ready the following year. Not long before their birthdays, they would shout down the speaker: 'It's my birthday in a few days time Mr Radio Man.'[27]

Positioning the presenters as aunts and uncles gave them authority and familiarity in the imaginary world collaboratively constructed by the radio and its young listeners. The presenters were jolly, unpredictable, friendly and amusing in ways that parental figures so often were not. The aunts and uncles could also enforce norms of behaviour on their young listeners and their unseen presence made them seem larger than life. But sometimes young listeners might get something of a jolt from their friendly radio aunts and uncles. A letter to Aunt Gwen shows the ways radio could be used as an extension of parental authority:

> Dad has just bought a new radio set and I would like to join your radio circle. I am six and a half years old. My birthday is on Feb 22nd. I listned [sic] to you and Uncle Toby and I liked it very much indeed. With love from Dick Falconer

At the foot of the page is a note in an adult hand:

> The above is a solo effort after a lot of work with pencil and rubber. Please reply next Saturday night as he will be away from home until then. Please tell him that he is quite good with the pencil but he must not write his name on the wallpaper. With many thanks, L.M. Falconer.[28]

Sadly, there is no record of how young Dick Falconer reacted when his beloved Aunt Gwen wished him a happy birthday with one breath and told him not to write on the wall with the next. The story probably became a family legend. Like Mary Matthews' appeal to 'Mr Radio Man', it illustrates the intimacy of radio and the hold it has over the imagination, especially the imaginations of the young.

• • •

Another popular form of radio entertainment between the wars was provided by sport broadcasts. Radio had been broadcasting sports news and commentaries since 1922, although some sport administrators were reluctant to let microphones near the events in case the practice drove down spectator numbers.[29] Live coverage of sports events requires imagination on the part of both the commentator and the listener. The commentator has to imagine the event as if it could not be seen, and the listener ideally imagines the event as if physically there. Allan Allardyce, sports announcer for 1YA, was commended as a 'born sports announcer' whose knowledge and commentating made 'listening in to his descriptions a genuine entertainment'.[30] But announcing could also make a sport event dull. One broadcast by 1YA of wrestling was described as 'a tame affair, slow announcing tended to break the narrative'.[31]

Radios were sometimes advertised as delivering an experience that was like being in a front-row seat.[32] This rhetoric was similar to that used to describe and sell gramophones and recordings, playing on the technology's fidelity to the original performance. But the claim for radio was not so much fidelity to sound as fidelity to experience. It was a way of saying that 'hearing is believing'.

Some seeing, however, could also enhance the hearing.[33] The *Auckland Star* was one newspaper that published diagrams as an aid to radio listeners in advance of rugby tests played during a 1930 tour of New Zealand by the British team. 'A Grandstand Seat by Radio', published in association with 1YA, was designed to help those unfamiliar with Auckland's Eden Park follow the location of the action during the test. The accompanying text claimed that the picture would help complete 'the mental image of the game' and cut out the need to 'stop and think', as the game would 'unroll itself naturally before the listener's eye' as the description was heard over the radio.[34] This added simple information to the listener's idea of the game without distracting from their listening. The sounds of the broadcast would in turn help the listener's visual imagination of the game for which the picture provided a contextual field. 'A Grandstand Seat' only had to be glanced at a few times to understand which part of the field was being described. The large sizes of the pictured players accentuated the proximity to the action that the radio delivered.

In conjunction with 1YA, the *Star* also published what it described as 'A Football Plan', but this may have proved more distracting. The idea behind it was that the radio announcer would call out the numbers of each square as he described the action moving back and forth across the field. This would have demanded more attention from a listener than the 'Grandstand Seat'. As each number was called, he or she would need to check where it was on the 'screen' to grasp where the action was on the field. Listeners could also keep track of the scores by noting the tries as they occurred.

The screen was an ingenious experimental attempt at helping a listener capture a sense of the flow and movement of the game through a combination of sounds and visual imagining. It points to the novelty of radio and how the medium was being developed and tried out by both broadcasters and listeners. It may or may not have helped listeners follow the game but it would have made a tangible souvenir in the same way a ticket could be kept as a reminder of a game.

Gordon Hutter, the announcer for the Auckland matches, was congratulated for his commentary of the 19 July game, but some found it too fast and too full of incident to take in.[35] Hutter seems to have taken this criticism on board, as his commentary of the 26 July game was praised in the press.[36] This newspaper story also contained an anecdote about a man who had spent the game in the crush of a packed stand. When he compared notes with a friend who had listened to it on the radio, he was so impressed by his friend's impressions and knowledge of the game that he declared, 'If you got all that by wireless, then I'm going to follow the next big game the same way.'

The 1930 British rugby tour took place at the height of the growing tension between the RBC and rugby and racing organisations over the

THE WORLD'S DIN

Live commentary on radio was just like being there. This advertisement appeared before the 1930 British Lions tour. Auckland Star, *13 June 1930*

This chart, published in the Star *each week, was designed to be used in conjunction with the radio rugby commentary. Commentators were instructed to call the names of the squares as the action moved around the field, and 'Enthusiasts … can mark on the figures denoting the square the initials of the team which makes each score.' History does not record the popularity of this initiative.* Auckland Star, *6 June 1930*

broadcast of those sports. Bookies and racing clubs were concerned that broadcasting was stealing their business and encouraging illegal gambling; the New Zealand Rugby Union and the provincial unions were worried about declining gate receipts as people stayed home to hear matches on the wireless.[37] This newspaper story may well have been a potshot in that particular struggle. By 1932, however, the dispute had been settled and sports broadcasts had become highly popular with radio listeners.[38] Games played overseas could also be followed through shortwave radio or on relayed transmissions by local stations.

On one level the broadcasts promoted national unity and feeling, just they do today. But these broadcast sports were played by males and their audience was mainly masculine, reflecting the central role of sports in the New Zealand ideal of manhood. Listening to sport was becoming a sociable pastime that might cement bonds of 'mateship', as well as family and friendship ties.

• • •

If sport was popular mainly with men, radio serials appealed to everyone. These were broadcast in New Zealand from 1931 and had an important place in the radio networks' schedules. They were imported in the form of 33⅓rpm discs, with one episode occupying one side of a disc. Serials such as 'Abroad with the Lockharts', 'The Japanese Houseboy', 'Eb and Zeb', 'Dad and Dave', 'Ambassadors of Melody Land' and 'Hy Wide and Handsome' became highly popular. Patrick Day estimates that they attracted up to 85 per cent of the listening audience during the later 1930s.[39] This popularity is reflected in the fond memories of listeners who could still name their favourite serials many years later. John Hill of Paeroa is a typical example. John and his brother Laurie bought a radio during the late 1930s. Hill vividly remembered his enjoyment of the serials 'Eb and Zeb' and 'Dad and Dave' almost 70 years later.[40]

The Australian radio serial 'Dad and Dave from Snake Gully' was based on the popular books by the writer Steele Rudd. The comedy was built around a family of Australian farmers, and ran from 1937 until 1953.[41] The rural humour of the Snake Gully adventures and pratfalls seemed to tap into the type of bucolic comedy that New Zealand writer Frank S. Anthony employed for his *Mark and Gus* stories from the 1920s and John Clarke later developed with his character 'Fred Dagg'.[42]

The fun of radio serials did not end when an episode finished. The serials stimulated the imagination and were a common source of gossip, jokes and stories.[43] But hearing the serials was important, and the sort of listening involved could be as focused as any the media theorist Rudolf Arnheim demanded. Betty Mules' mother 'used to love those Australian programs' and the children had to get out and leave her to listen in peace and quiet.[44] It is highly likely that time alone was just as important for Mrs Mules. Listening was as much a solitary as a sociable activity.

Despite the all-embracing model of Britishness that dominated much of New Zealand's cultural landscape during the inter-war years, Māori voices were heard from early on, as the *Radio Record* schedule shows. Performances by Māori cultural groups were a regular feature from 1928, beginning with the Wanganui Maori Concert Party performing as part of the 87th anniversary of the signing of the Treaty of Waitangi. The programme aroused much public interest and was repeated. J.F. Montague, an Auckland speech teacher, also began regular broadcasts on Māori pronunciation. Hare Hongi (Henry H. Stowell) of Ngā Puhi took over these in 1929.[45] This was not Māori broadcasting in any real sense: the programmes were made by Pākehā and were aimed at Pākehā. Montague's initial motivation was entirely didactic: 'Some of the common mispronunciations, many of them in fact, are atrocious, and it is time something was done about it.'[46] Just as speech teaching of English was normative and didactic, the broadcasting of Māori pronunciation was designed to be instructional and corrective rather than to raise awareness of the culture and language itself.

The performances by Māori were also not attempts to invigorate or arouse awareness of the culture. They represented the Pākehā view of Māori as naturally musical performers. Māori were presented as 'quaint natives' whose colourful stylings of European-based musical forms were part of New Zealand's national heritage, but were in no way an alternative to the performances of high- and middlebrow European music that radio stations habitually broadcast. An editorial in the *Radio Record* put it this way: 'Suitably enough, the pageants will be associated with the Maori race, whose poetic and musical genius, in association with their romantic past, promises entertaining evenings.'[47] The article refers to 'our romantic native race' in paternalistic tones typical of Pākehā attitudes towards Māori culture at this time.

Commercial recordings made of Māori musicians between the wars were highly popular and marked the formation of a distinctively New Zealand musical genre which featured Māori language and songs in European musical

Partygoers dressed as the popular radio celebrities Dad and Dave from Snake Gully in the late 1930s. Their costumes are modelled on those seen in the 1938 film spin-off Dad and Dave Come to Town. *In case anyone missed the point, they had also painted their names on their prop suitcase.* Alexander Turnbull Library, Wellington, Ref: PA1-o-1029-24-2

forms. The popularity of the radio broadcasts of Māori performers during the same period indicated the public interest in a sonic imagining of Māori culture which was non-threatening, sentimental and romanticised. Māori radio – that is, radio made by Māori for Māori listeners – emerged as a powerful force during the 1980s but the struggle was long and difficult.[48]

• • •

Māori were presented on radio between the wars to bolster the idea of New Zealanders as one people rather than as a diverse assemblage of groups and cultures whose interests and needs often diverged. This listening could be construed as a form of comfort for the Pākehā majority and a deafness to the fissures and divisions that ran through society.

The anniversary of the signing of the Treaty of Waitangi in 1840 was marked on radio in 1928 as an important moment in New Zealand's cultural life. It presented a version of the country's identity and history that stressed unity and peace, and gave Pākehā New Zealanders an account of themselves that was flattering and soothing.

But radio also allowed New Zealanders to listen out to the world and its events both joyful and solemn. In 1927 the BBC's experimental shortwave station at Chelmsford transmitted the sounds of Britain's Armistice Day ceremony to Australia and New Zealand.[49] Christchurch's station 3YA relayed the signal on medium wave so that people without shortwave receivers could hear it; Wellington's 2YA and Auckland's 1YA could not get the signal from Britain at all.[50] The subsequent development of the BBC Empire Service was followed with interest in local newspapers, and regular schedules began to be published in 1936.[51]

The sounds of Britain, still the 'home country' for many New Zealanders between the wars, were eagerly awaited. Shortwave broadcasts of events such as the Royal Jubilee of 1935 and the abdication of King Edward VIII in 1936 strengthened and extended New Zealanders' sense of the 'imagined community' of the British Empire in a direct way. As a child, Lorraine Russell loved the BBC news 'with the spine-tingling sound of Big Ben followed by the announcer's voice crackling over the distant airways'.[52] For Lorraine and the vast community of other 'listeners in' around the world, the sound of Big Ben's chimes became a metonym for Britain and, by implication, the Empire itself.

The last major imperial occasion that New Zealand listened to through radio in the inter-war years was the coronation of King George VI on 12

May 1937. It was the first coronation of an English monarch to be broadcast on radio, and was a huge undertaking involving hundreds of miles of cable, 58 microphones, 12 tons of equipment and 60 BBC engineers, 'those inside the Abbey doing their job in full morning dress'.[53] The event was marked by speeches, parades and banquets in New Zealand. Many people then stayed up late listening to the ceremonies being transmitted around the world on shortwave by the BBC Empire Service.[54] 'Listener', the Christchurch *Press*'s radio critic, playfully estimated that a million New Zealanders had tuned in to the live broadcast, then went on to paint a picture of the fortunate listener sitting comfortably at home, far from the crush of the crowds 'who could, with the exercise of a little imagination, bring the gorgeous scenes into our own sitting rooms'.[55] As ever, the exercise of the imagination was at the heart of radio listening.

•••

But imagination was not the only element involved in listening to the radio. Listening was embedded in a rich context of social and cultural practices. An extensive print culture provided listeners with guidance and information. The schedules printed in newspapers and the *Radio Record* offered a temporal framework for those who altered their daily routines to incorporate broadcasts into their leisure time. Celebrity culture was part of the fun of radio and regular columns in newspapers provided news, information and gossip about the people and institutions that made the broadcasts. There was also a rich and varied world of differing ideas about radio in the form of opinion pieces in newspapers and magazines, and letters to the editor. The print and pictures that went with radio were as much a part of listening as the sounds themselves.

Radio was heard in all sorts of social contexts. Listening could be solitary or communal. A rugby match or a popular serial might fill a room with people enjoying the social occasion as much as the broadcast itself. Families might be together with the radio on even if no one was paying much attention to it, and a broadcast such as Gwen's wedding or the King's coronation might be a communal experience for people who were not otherwise connected. Radio provided material for conversations and other social interactions. Talking about 'Dad and Dave' with friends or dressing up as the characters for a fancy-dress party were ways of connecting listening with everyday life that involved imagination and shared cultural knowledge. And these imaginative connections could extend much further beyond personal interactions.

Historian Peter Gibbons has written that the seas surrounding New Zealand were highways, not barriers. The goods of the world economic system travel along these highways to be consumed in New Zealand.[56] Analogously, the fields of electromagnetic radiation that carry radio waves are also types of highways to and from New Zealand. Between the wars, radio waves connected New Zealanders with increasingly grander imaginary communities, from the personal to the local level, to the national level and then to the global. With a radio, and a little imagination, New Zealanders were listening out to the world as well as in to themselves.

Part III | FILMS

CHAPTER 10

LIVING PICTURES

•••

In November 1895 just four years after Thomas Edison's improved phonograph had been demonstrated throughout New Zealand to great acclaim and wonder, another device billed as 'the latest of all the marvels which have emanated from Edison's wonderful brain' was displayed in Bartlett's photographic studio on Auckland's Queen Street.[1] The kinetoscope was a peep-show device that ran films of about one minute's duration.[2] The exhibitor, Alfred Whitehouse, charged patrons the somewhat hefty price of one shilling (children only 6d) to view four films, and also offered them the chance to record their voices on a phonograph.[3] The exhibition was extremely popular and Whitehouse kept it fresh by including new films throughout December. To keep the crowds coming, a kinetophone was added to the display in January 1896.

The kinetophone was simply a kinetoscope that had a phonograph player installed in its cabinet and earphones attached so that viewers could hear music as they watched a film.[4] The sound was not synchronised with the pictures in any way. Listeners at Whitehouse's exhibition watched a film loop of a dancer while hearing a recording of a band playing the 'Santiago Waltz'.[5] This was a quick fix rather than a solution to the problem of synchronising sound and vision – a situation Whitehouse tacitly acknowledged when he billed the kinetophone as 'the aggregation of scientific miracles'.[6] Thanks to reports of the invention of the kinetoscope in New Zealand newspapers, many of those who watched and listened in Bartlett's studio would have been aware that fusing sounds and images seamlessly together was Edison's goal.[7] According to one article, Edison's purpose was to combine the phonograph and the kinetoscope to the point where film of a pianist's hands would be in perfect synchronisation with the sounds of the phonograph: 'until he has accomplished this, he does not regard his work in this direction as done'.[8]

> **THE AGGREGATION OF SCIEN-TIFIC MARVELS,**
> **EDISON'S GREATEST ACHIEVEMENT,**
> **THE KINETOSCOPE,**
> **THE KINETOPHONE,**
> **THE PHONOGRAPH,**
> **THE GRAPHAPHONE,**
> **AT BARTLETT'S STUDIO.**
>
> **THE WONDERFUL KINETOPHONE**
> Exhibits the graceful Butterfly Dance by Anna Belle to the music of the Band playing the Santiago Waltz. Must be seen to be realised.
> A. H. WHITEHOUSE, Exhibitor,
> Bartlett's Studio.
> **THE ONLY EXHIBITION OF THE KIND IN NEW ZEALAND.**
> See them at once, as our season closes shortly.
> N.B.—Would you like to hear your own voice. You can get it recorded on our loud-speaking Phonograph.

Moving pictures, along with talking machines and wireless, were yet another marvel of science and progress. Whitehouse's advertising promised all manner of modern technological sensations. NZ Observer and Free Lance, *25 January 1896, 16*

Edison never did develop a usable technology that synchronised sound and vision; New Zealanders had to wait until 1929 to see and hear their first feature-length talkie.[9] But until then local film viewers were exposed to a rich variety of sounds as they watched the flickering images projected in theatres, churches, town halls, cinemas and wherever else a projector might be set up. It is a cliché to write that it is a cliché that silent film was never silent. However, both clichés are true. Film was never a silent medium, although its silences might be just as important as its sounds at various times and places.[10] But the history of New Zealand cinema is something of a mime show in that very little attention has been paid to the sounds of cinema in general, let alone those of the silent era. As an example, the latest general survey of New Zealand cinema makes only passing references to audio, and these are mainly about the arrival of synchronised sound (the talkies) during the late 1920s.[11]

The rise of the field of Sound Studies over the last 20 years, however, has seen scholars across a variety of disciplines turn their attention to the important roles of audio in history, cultures and media.[12] It is no longer possible to describe cinema or television as 'visual media' and simply ignore the vital roles that sounds play in their construction and reception. They have always been *audio*-visual media. It is as if many writers (reviewers, critics, academics and so on) watch television, films and games with the sound turned off, because precious few of them discuss the audio that is inextricably part of the experience.[13] The screens that surround us are as much audio devices as visual devices, and we must take this into account to fully understand both old and new media.

New Zealanders enjoyed the rich soundscapes surrounding silent films during the years between Whitehouse's exhibitions and the advent of synchronised sound. The variety of sounds heard with silent films ran against the standardising of experience associated with the technologies of modernity. Film exhibitors used a wide range of techniques and technologies to generate sounds that filled the acoustic spaces around the films. These included orchestras, lecturers, pianos, actors, audience members, wind machines and anything else that could make sound.[14] Improvisation was often central to the sounds of the silents.

The range and variety of these experiences complicates the idea of cultural modernity as a series of mass-produced and standardised experiences or a 'culture industry'.[15] The silent films available to New Zealanders were mass-produced copies of images, but the sounds heard with these films were anything but.

It was one experience to watch a film such as *Quo Vadis* (1913) in a plush theatre with an orchestra and choir performing the especially composed score that went with it. It was quite another to see the same film in a school hall with a pianist improvising music to suit the action onscreen. Ignoring the acoustic experience of historical silent cinema viewing gives us only half the picture. The richness and complexity of silent cinema runs counter to the popular idea of film history as triumphant progress from flawed, primitive, silent beginnings to a sophisticated and perfected present, with the advent of the talkies positioned as a revolutionary lynchpin. This chapter and the next are attempts to restore the sounds of silence and recover some of the sonic delights that went with the flicks.

• • •

The acoustic practices of New Zealand silent cinema overlapped and differed in time and space. It is difficult to structure the material into a strictly chronological arrangement. Such an arrangement would not reflect the complex and interlinked methods of sonic production and reception that characterised the silent cinema era in New Zealand. These sounds included those made by orchestras, pianos, phonographs, coconut shells, drum kits, film audience members, lecturers, and the machinery of the cinema technology itself. This chapter discusses how human voices and recorded sounds were integral parts of early cinema.

The human voice played varied and surprising roles in silent films — from bare description to poetic recitation — from the first appearance of moving pictures in New Zealand. The first film exhibitors used their voices to describe both the equipment and the action on the screen. Neither was seen as self-explanatory. The lecturer (like the phonograph demonstrator) might expound on the technology being used, and also explain what the moving pictures depicted. During his 1896 exhibitions, J.F. MacMahon 'announced the subjects' of the films while a Mr J. Margery operated the equipment.[16] Newspaper accounts of the technology, which were often part of the exhibitions' publicity, seem to have been common in the very early years of film exhibitions, and served to familiarise the curious with the details of the equipment.[17]

Women as well as men worked with the new technology of moving pictures. This had not been the case with the earliest phonograph exhibitions. The moving pictures were shown in theatrical contexts where women were established as performers, so they became another element of stage work, the preserve of versatile entertainers. The Christchurch Kinematograph Syndicate toured New Zealand through 1899–1900. Typical of the itinerant film shows of the time, it featured several singers, comedians, lantern slides and musicians, as well as a variety of kinematograph films including scenes of the South African War. This was a varied night of entertainment, and the films were just one part of it. The performers took on various roles. Harry Baxter, a well-known and popular singer, 'announced' and explained the films.[18] Sometimes he combined both roles of announcer and singer as the film played. One reviewer commented on the scene of a troopship leaving for war, 'the effect of which is enhanced by Mr Baxter's singing of an appropriate chorus while the picture moves across the screen'.[19] Baxter also sang a brace of songs between the two film sessions of the evening; these too were warmly received.

In 1900 the Happy St. Georges Company showed kinematograph films including '4000 ft. of animated pictures of the *Passion Play*, with a young lady lecturess', and also offered the traditional repertoire of singers, lantern slides, dancers and variety turns.[20] This 'young lady lecturess' was Nena Manning, and she was an able singer. Her repertoire included 'Oro Pro Nobis', 'The Lost Chord' and the 'pathetic song', 'Little Hero', all of which were illustrated with slides and were well received.[21] As versatile a performer as Harry Baxter, she also described the film of the *Oberammergau Passion Play* that was the company's drawcard. Her descriptions seem to have been as appreciated as her singing. One reviewer noted that her 'distinct delivery and grasp of elocution materially added to the pleasure and interest in the entertainment'.[22] Another described Manning as 'a lucid lecturess'.[23]

The announcing roles of Baxter and Manning raise some interesting questions about what they said, how much they said and what their words meant to the audience. The sequences of the war films and the *Passion Play* were published in newspapers, and the broad outlines of what the films showed may well have been known to the audience in advance, particularly the events of the *Passion Play*.[24] Did the speakers say only a few brief phrases that gave such details as names, locations or short descriptions? Manning was described as prefacing each view with 'explanatory remarks', which implies that she did not talk all through the pictures but rather set the scene for the viewers so that they could follow the action without distraction.[25] The South African War films often provoked noisy audience responses. Faced with such reactions, speakers such as Baxter may well have struggled to be heard and perhaps made only brief remarks. What was said or left unsaid may have varied from audience to audience.

While some film announcers may have said little, some perhaps said too much. The British Biograph Company toured films in New Zealand during the early 1900s after successful seasons in London and Australia.[26] Their shows included scenes of England, comedy pieces and striking images of the South African War. There seems to have been no description or announcement of the pictures. This certainly pleased one viewer, who praised the realism of the films, the sharpness and size of the images, and the 'absence of the showman's lecture, which is usually so full of egregious blunders and misstatements that it grates upon the nerves'.[27] Instead of spoken introductions, titles and descriptions were displayed before each film. Much to this viewer's relief, 'not a word was spoken'. Incidental music before each film was provided 'by an accomplished musician', presumably on a piano, and each tune was 'grave or gay, martial or dismal, according to the picture'.[28] The combination of titles and music seems to have performed the role of the spoken words supplied by performers such as Baxter and Manning, imparting information and setting a mood for each film. This would have been different at each performance, just as the theatre experience with actors varies with each performance. Different audiences would see the same images at each screening, but the sounds that went with these scenes were varied and open ended in ways that were drowned out by the recorded soundtracks of synchronised films.

The rise of fictional narrative films during the early 1900s called for more complex and, at times, continuous commentaries from film announcers to match the more complex plot structures. These longer films, both comic and serious, went beyond the 'emphasis on display rather than storytelling' that characterised the early 'cinema of attractions'.[29] The rise in popularity of such so-called 'picture dramas' in New Zealand, as elsewhere, was noted by one writer in connection with the activities of the Australian actor Alfred Boothman, who was well known in New Zealand in the early 1910s.[30] One of his main activities was as a film announcer, but he seems to have gone far beyond the short scene-setting that Manning and Baxter had provided. The Australian-made film *For the Term of His Natural Life*, a dramatisation of the book by Marcus Clarke, was very popular in New Zealand during 1910. At about one hour in duration and with many scene changes and plot elements, the picture was praised by reviewers for its stirring cinematography and ambitious scale.[31] Boothman's role in this production was to narrate the events that unfolded on screen.[32] As one reviewer wrote, Boothman 'describ[ed] the main features of the drama'.[33] Boothman carried on narrating films until 1914. He was described as 'speaking the dialogue' for the 'stirring' film *The Kelly Gang* in 1910. A reviewer commended his 'clear, expressive voice' and noted that his 'splendid oratorical efforts' were admired by the audience; 'to have the tragic story of the Kellys explained as the pictures were thrown on the screen was to make one feel that the stirring events being portrayed were being enacted in the flesh at the moment'.[34] For this reviewer at least, Boothman's spoken commentary added to the images on the screen. But just what was added is unclear.

The words used in conjunction with Boothman's performance – 'describing', 'explaining', 'telling' – conceal as much as they reveal. Did he describe scenes and actions, or did he say things that might have been said by the characters? Did he do both? What did the 'explaining' of the pictures consist of, and how did his 'oratorical effects' contribute to the realism felt by the *Argus* reviewer? It is probable that Boothman worked from scripts, but without these scripts it is impossible to recover just what he, or other such announcers and actors, may have said. We do not know whether they improvised to suit local conditions or if different speakers used the same scripts. The range of choices indicates the variety of possible experiences with early cinema. The moving images looked the same wherever they were screened, but there was nothing standardised about the sounds that went with them.

One later performer in New Zealand who did leave a script for the historian to consult was Barrie Marschel. Educated at Eton and Cambridge, Marschel practised law in Australia before becoming involved with theatre from vaudeville to Shakespeare, and even working with the famous actor Lily Langtry in the US. He came to New Zealand in 1905 and taught elocution in Dunedin as well as touring the country with theatre companies. Marschel became involved with the new cinema industry in 1913 as an advertising manager for New Zealand Motion Picture Supplies. After World War One he established British Dominion Films Ltd, a trans-Tasman company that eventually became the local part of Twentieth Century Fox. He died in 1940 after a career that crossed from traditional theatre to the new art form of cinema.[35]

Marschel became famous in New Zealand on the screen rather than the stage when his 15-stanza poem about Gallipoli, 'The Kid from Timaru', already popularised in theatres as a recitation piece, was filmed in 1917.[36] This exciting, sentimental and patriotic one-reel feature showed the events of the poem, and Marschel toured the country reciting the verse in time with the action on screen. Marschel had some experience as a film lecturer, having provided 'an explanatory lecture' to a 1914 film biography (an early 'biopic') of Queen Victoria.[37] The *Kid from Timaru* film was described as a 'poem picturised'.[38] It was in some ways an extension of the theatrical 'one-hander' shows during which a solo actor presented the piece.[39] In this case, however, Marschel presented an entertainment using the popular modern technology of cinema to create a suitable backdrop for his recitation.

Marschel claimed to have written the poem after reading a letter that described the exploits of a young soldier from Timaru. Some have identified the 'Kid' ('Kidd' in the poem) as being based on James Hagerty, a famous boxer originally from Timaru who died at Gallipoli on 27 August 1915, but this is now disputed.[40] The poem uses a ballad structure and simple rhymes that make it easy to remember and understand when heard rather than read:

> A rousin' cheer, that split the sky, went boundin' through the air,
> We vowed when we struck Gaba they'd know that we were there.
> We swore for king and country our very best to do,
> Kidd swore for king and country, but added – Timaru.
>
> We faced 'Loose Hell', as scrunching o'er the sand we scaled the cliff,
> While Turkish snipers' rifles mowed men down at every whiff;

> No fellows stopped to count the cost as up the bank they flew,
> And level, with the foremost ran young Kidd from Timaru.
>
> And when the fight was over and each had done his part,
> And felt a man and soldier, with aching eye and heart,
> I searched among the wounded for the fellow that I knew,
> I turned one over on the sand – 'twas Kidd from Timaru.
>
> He'd carried in his Captain, almost dying through the wrack,
> Of smoke and fire of battle; but just as he'd got back,
> A Turkish sniper 'pink'd' him but the bullet went clean through,
> And when he's well they'll hear again from Kidd of Timaru.[41]

The impact of the war and the sentimental and heroic nature of Marschel's story ensured the popularity of the show. His feature-length vocal performance was an integral part of the experience. This type of silent cinema required close listening from its audience members.

Some spoken film shows were intended to do more than entertain and inform. As films became adjuncts to the already familiar lantern slides, moving images became an important part of the lecturing involved in popular education and self-improvement.[42] Lecturers often used films to both emphasise their messages and attract listeners. The New Zealand Salvation Army, for example, was quick to take up films as part of their struggle against sin in general, and drunken sin in particular. Joseph Perry, the Australian-based pioneer cinematographer and major in the Salvation Army, made and toured films throughout New Zealand during the early 1900s in the service of the 'Good Fight'.[43] These shows combined band music, lantern slides and moving pictures.[44] Viewers seemed to admire the variety of pictures shown, as well as the music from the highly drilled musicians.[45] In fact, some may have felt the moving pictures to be far more entertaining than the lectures that accompanied them. While many may have heard such lectures as uplifting adjuncts to moving pictures, some may have shared the opinion of the British Biograph viewer that the words grated on the nerves.

It is difficult to know just how lecturers incorporated films into their presentations. Did they speak continuously over the films? Were the films shown before, after or during their speeches? Magic-lantern slides could be paused as long as needed to make a point, but film images continuously flowed past the audience. The moving pictures could not be paused, as the

A poster for Barrie Marschel's popular film The Kid from Timaru, *which played to full houses the length of New Zealand until early 1919. The footage of the landing at Gallipoli was an enactment shot in Australia.* Alexander Turnbull Library, Wellington, Ref: Eph-A-CINEMA-1918-01

This cartoon is from a page of satirical cartoons entitled 'How we kept Trafalgar day in Auckland'. Now little observed publically, Trafalgar Day used to commemorate Lord Nelson's famous 1805 victory over the French navy. The worthy on the stage is about to deliver a patriotic 'speech on Trafalgar', but the sea cadets had been expecting 'moving pictures'. New Zealand Observer, *28 October 1905, 12*

film might melt or burst into flame if stopped for too long. This made long explanations about particular scenes or moments very difficult to manage while the film played.

The role of lecturers and actors in narrating films in New Zealand and elsewhere seems to have died away by the early 1910s, although the tradition became well established in other countries.[46] This decline may be attributed to the rise of intertitles and the increasingly sophisticated visual narratives employed by film makers.[47] But an equally important factor was the music that was used to underpin, amplify and comment on the meanings of the pictures.

The potential power of music in silent cinema was hinted at by Jean-Paul Sartre when he wrote of his childhood silent-film heroes: 'they weren't mute since they knew how to make themselves understood. We communicated by means of music; it was the sound of their inner lives.'[48] It is hard to know if many New Zealanders who went to silent films felt the power of the music as strongly as Sartre claims he did. The music of silent films served many functions at different times and places, and was produced and heard in a bewildering variety of ways. These might include large orchestras in modern, richly plush cinemas, a battered upright piano in a remote town hall, or singing by audience members in a vaudeville theatre. Sartre's comment illustrates just one reaction to the music of silent cinema, and a retrospective one at that. But his words point to the importance of music as part of the experience of silent cinema. Films became standardised in terms of duration and genre as the cinema industry grew, but the musical sounds of films remained diverse and unpredictable.[49]

• • •

Recorded music, whether on wax cylinder, shellac 78, vinyl LP, magnetic tape, cassette or as a computer file, has always been part of cinema-going in New Zealand. Sometimes the recordings have been as important as the films played with them. At other times, the recordings have been little more than background noise or a means to fill in time. Sometimes this was the case at the same screening. It is hard to generalise about the use of technologies such as phonographs and gramophones in cinema before the late 1920s, but some broad patterns can be outlined.

Phonographs seem to have been an important part of the shows staged by most film exhibitors during the late 1890s and early 1900s. As dedicated cinema venues became common in most cities and towns between 1910 and 1920, recorded sounds became less central to cinema viewing, and were used more as a means of filling in time during intervals or while film reels were changed. Many cinema theatres used live musicians to supply all the music during a film showing rather than relying on recorded sounds. In more isolated areas, where mobile cinemas were common, phonographs and gramophones were convenient and portable ways of supplying music that might be an added attraction to the films themselves. But this also varied; a local musician could supply music on a piano that might in turn be in addition to recordings. The use of canned music by no means meant that film experiences were also pre-packaged and identical.

• • •

Most of the touring film exhibitors during the late 1890s and early 1900s used phonographs. A.H. Whitehouse's first exhibitions of moving pictures with the kinetoscope and kinetophone in 1895 also included phonograph demonstrations. The many tours of films that he made from 1896 until his retirement in 1908 always included phonographs and recordings.[50] Whitehouse had exhibited a phonograph in 1894 and claimed that the experience he gained in the course of that tour 'created a desire to bring out Edison's later invention, the kinetoscope'.[51] These two technologies, one doing for light what the other did for sound, were inextricably linked from their first appearances in New Zealand.

The credit for shooting the first moving pictures in New Zealand is generally given to Whitehouse for his images of the opening of the Auckland Exhibition in 1898.[52] He went on to make many more short features and was a prominent roving film exhibitor. In accounts of New Zealand film history, Whitehouse's phonographic activities tend to be overlooked in favour of his film-making and touring.[53] However, both making and playing recordings seem to have been important aspects of his tours and exhibitions. So strong was the association of Whitehouse with recorded sound that he was incorrectly described as 'the first man to introduce the phonograph into the Dominion' in a 1949 obituary.[54]

Whitehouse seems to have played films and sound recordings separately, a pattern repeated by many other exhibitors of these new technologies.[55] Joseph MacMahon, for example, toured New Zealand with a cinematographe machine during 1896 and 1897. He presented an organised and varied show that resembled those of the carefully prepared phonograph tour by Archibald. MacMahon's show eventually grew to include 25 films, among them depictions of the internationally famous strong man Eugen Sandow, a cockfight, boxing cats, and a re-enactment of the execution of Mary Queen of Scots.[56] Reviewers described these exhibitions as 'supplemented' and 'diversified' by phonograph displays.[57] MacMahon seems to have played the phonograph in an interval between sets of films.[58]

Several months after MacMahon's tour ended in 1897, Neilson & Sons were showing films in Auckland. A review of one of their 'Phonoscopic Concerts' discussed the films and sound recordings as discrete parts of the display: 'A large number of items were given out on the phonograph which was a splendid machine. All the best items on both machines had to be repeated in some cases several times.'[59] One month later, Taylor & Jones gave a film and phonograph display in Ōtāhuhu's Public Hall 'consisting of some good cinematographe pictures and an exhibition of Edison's loud-speaking concert phonograph'.[60] These also seem to have been played separately. An account from 1898 of a film show stated that 'each picture was alternated with a record from a loud sounding phonograph'.[61] In 1900, the Boer War

Biograph Company advertised that their film shows at Auckland's Opera House featured live music and recordings 'rendered on the new Polyphone, the most perfect talking machine in the colony'.[62] During the same year, the Fisk Jubilee Singers brought phonographs and films on one of their many tours of New Zealand.[63] Their modern phonograph attracted much comment, but its selections seem to have been played as discrete parts of the show.[64] Four years later, the Biograph Company entertained Tauranga with films of the Russo–Japanese War, comic scenes and illusions. A large phonograph also 'gave musical selections at intervals'.[65]

In 1906 another company used the name 'British Biograph' to tour films and phonographs around New Zealand. Rowland Chubb's British Biograph Company showed films of the damage caused by the San Francisco earthquake.[66] Chubb had arrived in Auckland with £2, a camera and a 'small moving picture set' in early 1903, and set up the British Photographic Studio on the corner of Pitt Street and Karangahape Road. He began showing films at the Karangahape Road Forester's Hall in 1905. Chubb also used other theatres but claimed that he was the first to 'attempt to establish permanent moving picture entertainment in Auckland'.[67]

This statement may have to be taken with a grain of salt, as Chubb seems to have been a little free and easy with the facts. With his films of the San Francisco earthquake, Chubb gave lectures full of lurid details based on his eye-witness account. He stated that more than 400,000 people had died, and that at one point 400 of the injured were 'mercifully chloroformed' as they could not be evacuated before the fire reached them.[68] The actual death toll was nearer 3000 and there are no records of mass euthanasia. There was another problem with Chubb's 'eye-witness accounts'. It was pointed out in the press that Chubb had left Auckland on 4 May, so he 'could have hardly witnessed the turmoil on the 18th April and following days. Clairvoyance might explain the puzzle. Either that, or vivid imagination.'[69] The use of the title 'British Biograph' was also misleading. Chubb appears to have no connection with the prestigious British film company that had toured New Zealand to great acclaim in 1900.[70] But his movies were viewed with interest by New Zealanders who enjoyed the titillating and blood-curdling (if somewhat exaggerated) account eloquently delivered by Chubb. The phonograph would have provided variation during the show and also allowed time to set up the next film. Chubb's shows were not at all death, doom and destruction. Along with the devastation of San Francisco, 'a number of fine views of Honolulu were shown and described'.[71]

Rowland Chubb (second from right) cashed in on the 'British Biograph' brand when he toured films of the San Francisco earthquake aftermath during 1906. This equipment ('gas tank, Edison cylinder phonograph and laced cinematograph') was typically used by exhibitors during the early twentieth century. Dave McWilliams, Auckland's Film History 1896–1941, 8

These examples show how film exhibitors developed formats and strategies for their displays that featured audio technologies as separate but equal parts of the entertainment. The recordings seem to have been most often played in sets before, between or after film showings, but usually not during them.

The timing of when recordings were played may have given performers time to set up the next stage of the entertainment. Or it may have been that the operator of the film and sound technologies could only do one thing at a time: these machines were quite difficult to work properly. Nonetheless, it is clear that both audiences and performers expected film to be accompanied by sound, even if this simultaneity was a long way from synchrony. The specialised technologies that played sound recordings and films at the same time were semi-experimental devices and did not attain lasting commercial success, even though the machines impressed some viewers. The kinetophone of 1895 is an early example. Another such technology, the Gaumont Chronophone, toured New Zealand in 1906 and moved one viewer to write that the effect was 'highly pleasing' and that 'for all the songs, the sound kept pace with the singer's lips as seen on the screen'.[72]

The effect of the Chronophone pleased many who witnessed and heard it.[73] In Dunedin it played to a 'very large and appreciative audience'. The close match between the songs and the 'facial movements of the performers' was highly successful during the varied musical items. The operatic selections led one viewer to predict that 'one of these days no doubt a whole opera may be seen and listened to under similar conditions'.[74]

Another device that matched music with images toured New Zealand in 1909. The Cinephone had already become popular in Britain and America. A local description summed up its appearance and possibilities:

> A gramophone projects its glittering brass funnel through the picture screen, and, while the film shows some music-hall artists at the other end of the world doing their 'turn', the 'phone brings you within range of their voices or of the strains to which they are dancing. Thus you may go down to Fullers' any night and witness a turn at the Empire in London. Bye and bye you won't need to cross the seas in order to do London. You'll be able to turn it on without leaving Port Nick [Port Nicholson].[75]

But 'Footlight' went on to add that the tone was neither full nor melodious and that the Cinephone was more an interesting auxiliary to the pictures. Others also found the sound inadequate.[76] New Zealanders might be able to see the music-hall greats of London's Empire 'without leaving Port Nick' but hearing them was still difficult, at least at the pictures.

Among the Cinephone attractions that particularly appealed to New Zealand audiences were the short films of the much-loved singer Harry Lauder.[77] But as with the Chronophone and other early sound synchronisation devices, these films were constrained by the storage capacities of the cylinders and discs they relied on. With only about four to six minutes of sound available, the devices could play only short films. Popular songs, brief vaudeville routines and short comedy sequences were suitable, but at a time when film durations were increasing and their plots and characterisations becoming more complex, such a limited amount of acoustic playing time meant the machines were entertaining diversions and curiosities rather than viable ways of synchronising recorded sound and vision.

The roles of recorded sound with films changed as purpose-built cinemas became common in New Zealand during the 1910s.[78] Here phonographs and gramophones appear to have supplied music as background noise while patrons entered or left the venue and while the projectionists prepared their next selections. Douglas Lawson recalled going to films at a cinema in Napier during the 1920s that was owned by a classmate's father, where 'the son's job was to play recent recordings on a big high-tech electric gramophone before screenings'.[79] Squire Speedy remembered the use of records as supplements to the films shown at Milford's Picturedrome cinema during the 1920s, when the gramophone provided incidental music during intervals.[80] Records were sometimes played at Paeroa's Gaiety Theatre during the early 1920s, as Eileen Mathieson recalled in 1977. When a piano player could not be found, 'a gramophone was resorted to and those in charge of it … would get so interested in the picture that the "winding" was forgotten'.[81]

Recorded sounds were to become central to the experience of cinema during the late 1920s with the advent of synchronised sound. But for some, recordings had outworn their welcome long before this. When Valentine's Picture Company visited Tauranga in 1907, one reporter was relieved to note that the 'monotony of pictures' was to be broken up by songs, sketches and instrumental interludes 'which will be a relief after shows that go through the evening with nothing but pictures and a phonograph'.[82] Variety was the thing in the film game, and sonic variety was provided by live musicians.

CHAPTER 11

ORCHESTRATED PICTURES

•••

Before the arrival of the talkies during the late 1920s, cinema goers were treated to a wide variety of sounds down at the flicks. Some of these were canned in the sense that they were recordings. Others were scripted (or at least prepared) words spoken by actors and lecturers. But the new medium required more than just scripted words and several-minutes-a-side recorded music to realise its expressive possibilities more fully. Sound effects and music made by people could be finely coordinated with the ever-changing scenes and action of the movies. Like jazz players, the musicians of the silent films responded to the audiences, the moods and their surroundings to often create unique soundtracks which heightened the viewers' emotional reactions to the films. At times, too, the noises (and the silences) of the audience members were important aspects of cinema's soundscapes. These all acted together in different ways in different places. To simplify this bewildering polyphony, this chapter examines sound effects, then music, and finishes with the most important element of cinema: the audience.

•••

Sound effects straddled the line between music and noise, and were used to add realism to the pictures. They could be generated by musical instrument such as drums or violins, but were also produced by devices that were non-musical. They could be a vital part of the cinematic experience and were probably very common when films were shown in theatres in which a variety of sound effects devices was readily available.

In early 1905 the Williamson's Bio-Tableau film company toured New Zealand with vivid footage from the Russo–Japanese War. The films included scenes of Russian troops on the move, the artillery at the Battle of Liaoyang and the siege of Port Arthur. Along with these dramatic scenes, there were also films of a German motor derby, the 1904 Melbourne Cup, a wrestling match, a bull fight in Seville, the 1904 Great Fire of Toronto and 'others too numerous to mention'.[1] The advertising and descriptions also mentioned the sound effects that went with the films:

> To enhance the interest and realism of the show the management have introduced a clever and telling series of realistic effects. As the barrel-shaped motor cars are sweeping along the road in clouds of dust, the hissing noise of the engines is pleasantly produced, and the regulated thud of marching troops will invest the war views with additional realism. The passing of cavalry, the roar of the guns, the rattle of artillery, all are faithfully and extraordinarily portrayed.[2]

The word 'portrayed' is used in the sense of a picture, but also to mean that the 'realistic effects' will live up to the description. And for some they did. One reviewer commented on the 'realism heightened to the top notch by the faithful reproduction of the sounds which ought to accompany the actions in the pictures'.[3] This included the laughter of the viewers at an exhibition, and a scene showing the demolition of a fire-damaged building in which 'the noise of an explosion crashes in your ear and simultaneously the eye beholds a lofty mass of masonry fall to the earth'. All this as well the noise of the artillery, cavalry, cars and so on made for a noisy night at the cinema. But these sounds were different at different venues, all of which had their own sound-effects devices. These could be complex.

This complexity is shown in a lengthy and fascinating account of sound-effects techniques by a reporter who witnessed the behind-the-scenes and screen activities at Dunedin's Alhambra Theatre, where the Royal Pictures Syndicate ran a season of films in 1908. The sound-effects operator was stationed behind the projection screen and used a variety of tools to 'give the necessary air of reality to the rapidly passing episodes depicted on the screen'.[4] These included tins of broken glass to suggest breaking crockery 'in the comic pictures'; a tray of small shot (or lightly rubbed sheets of sandpaper) to make the sounds of waves; iron wire rubbed on a metal plate for a locomotive; hinged wooden laths for gun shots or whip cracks; a heavy wheelbarrow for 'wheeled traffic'; coconut-shell halves for horse hooves; and many others. These were standard theatrical effects but were now used to give 'the aspect of reality to what would otherwise be but cleverly pictured pantomime', and were described as 'a most important part of the modern cinematograph exhibition'. At times the busy sound-effects operator seems also to have supplied speech for the films as 'the man who produces the effects, spoken and other, which give the necessary air of reality to the rapidly passing episodes depicted on the screen. With eyes fixed upon it he hurtles out appropriate dialogue in surprisingly facile style.'[5] But there is no mention of a script. Did he simply improvise or did he prepare some lines? A number of the films shown during the season at the Alhambra were nature films and some of the spoken words may have been informative as well as dramatic or comical. But the audience members certainly heard a great many varied sounds.

Not everyone agreed that such effects enhanced the realism of the films, and there was much debate both in New Zealand and elsewhere about whether such effects were needed or appropriate.[6] The film *Living London* attracted much attention when it was shown in New Zealand in 1906. Showing a day in the life of the city most New Zealanders regarded as the centre of their Empire, it was bound to be popular. The sound effects used to enhance the film attracted some comment. One viewer in Whanganui wrote that the sounds were 'most natural' and helped the audience realise 'what actual life in the metropolis is'.[7] Perhaps the effects operator had heeded the comments made a week earlier by a disgruntled Wellington viewer who enjoyed the pictures but complained that 'the labourers behind the scenes' worked too strenuously to produce the sound effects. Their range was limited and 'did not do the excellent pictures justice'.[8] These two responses illustrate the opposite poles of the contemporary debates about sound effects and silent films. To some they sounded like fakes: to others they were realistic and enhanced the films.[9]

This debate about sound effects faded out in the age of dedicated cinemas, pianos and orchestras. But musicians, mainly percussionists, still drew on a range of noise-makers to add to the films. Bert Vinsen played drums in a Taranaki theatre orchestra during the 1920s and used cow and sleigh bells, motor horns and coconut shells, along with pistols 'that fired blanks to pep up the Westerns'.[10] Such literal representations of the screen action may have suited comedies and action pieces, but live music was more suitable for the emotionally complex and character-driven dramas that became popular in the dedicated cinema venues after the 1910s. A 1916 article about the mechanics of the movies described an Auckland cinema: the only reference to sound was that of 'the tinkle of a piano'.[11] Sound effects were a useful part of the sounds of the silents. But it was music, the most expressive of the arts, that made the strongest emotional connections between the viewers and the images.

•••

Prior to the spread of dedicated movie theatres, film companies usually combined with variety and vaudeville acts, and worked the established theatre circuit. Perhaps the most famous of the variety and film combinations was West's Pictures and the Brescians. Thomas James West's Pictures toured New Zealand between 1905 and 1908 with the Brescians, a musical troupe founded in Britain by Henry Hayward, who later became a prominent cinema theatre owner and operator in New Zealand.[12] Having cut its teeth touring Scotland and England, the West's and Brescians' show was slick and professional by the time it arrived in this part of the world. With no pauses or intervals, the parts of the show fitted seamlessly together, creating an evening's entertainment that smoothly combined traditional songs and musical items with the latest cinematic technology. The shows were organised, as one reviewer had it, like a sandwich, with West's Pictures being the bread and the Brescians the jam, and that 'right throughout, crust and crumb, it is of first-class quality'.[13]

The evening began with film scenes of King Edward, during which the Brescians Chorus performed 'God Save the King'.[14] Some travel films of Italy and Spain, along with several comedies, filled out this first 'crust' of the show. The jam in the middle was the Brescians with a varied programme of songs and instrumental pieces. The last crust of the evening's entertainment consisted of more films. In the view of one later New Zealand commentator, these film shows 'marked the beginning of organised picture-screening and of the picture entertainment as we know it'.[15]

The silence of the silent films meant that language was no barrier; intertitles in any language could easily be edited into films.[16] Among the international films that circulated through New Zealand were some innovative works. The Brescians toured with *An Impossible Voyage* (*Voyage à travers l'impossible*) by French director Georges Méliès. Méliès is recognised as an important figure for his pioneering use of special-effects techniques that drew on his experience as a stage illusionist.[17] *An Impossible Voyage*, an early science fiction film, was 20 minutes long, nearly five times the average length of contemporary films. It was a spectacular fantasy with trick photography, subtle colour techniques, and an outlandish story involving a journey to the sun, along with trains, submarines and balloons. This film, Méliès' sequel to his 1902 *A Trip to the Moon* (*Le voyage dans la lune*), greatly impressed its New Zealand viewers.[18] Just 29 years later, journalist and writer Robin Hyde incorrectly described *Impossible Voyage* as 'the first picture ever shown in New Zealand', perhaps indicating the impression it made on its viewers.[19]

Others seemed to find the pictures more interesting than the Brescians' musical abilities. As a boy, politician and writer John A. Lee saw the West's and Brescians' show in Invercargill in 1905 and later recalled that 'the music of the Brescians was beyond me. I could not rise beyond a music hall song or a revival hymn, and waited eagerly for the picture'.[20] The music for these shows seems to have been well separated from the films, apart from the opening scenes of King Edward. Did the audience sometimes accompany the Brescians in 'God Save the King'? This remains a tantalising possibility.

It is unclear whether music was played as the films were screened. Given that there were no intervals, the instrumentalists may have been on stage throughout the show even as the singers came and went. A later source mentions the use of a commentator to explain the films, and this may have precluded music.[21] One reviewer described the show as 'a veritable marriage of the senses' in that 'pictorial symphonies captivate the eye whilst the musical harmonies charm the ear'.[22] The words 'marriage' and 'whilst' do not necessarily mean that the pictures and music happened simultaneously, but the comment is suggestive of the overall effect of the West's and Brescians' slick showmanship. Certainly, by the time West's Pictures returned to Wellington Town Hall in 1908, this time without the Brescians, the music supplied by De Groen's Vice-regal Orchestra included 'up-to-date and popular musical selections' to accompany the films.

The development of dedicated cinema theatres invigorated New Zealand's musical life by providing venues where not only films but also a large variety of music could be heard. Typical of these venues was the Empire Theatre in Auckland, which opened in 1911 as one of Henry Hayward's Picture Enterprise's chain of theatres. Like most of the cinemas built in New Zealand during the silent era, its interior was modelled on traditional music theatres where the music or voices were not amplified. Theatres such as the Empire had to be modified to meet the acoustic demands of the electronically amplified sounds of the synchronised sound films.[23] The orchestra pit was placed in the space between the screen and the first row of seats. The musicians had their backs to the screen but the conductor could see all the action and so direct the players accordingly. The piano player, if required, would also have been placed where he or she could see the screen. The musicians needed quick reflexes to keep up with the rapid changes of onscreen mood, action and setting.

The interior of Auckland's Empire Theatre c. 1914. The orchestra pit can be clearly seen below the screen with some sheets of music left on the stands. The inset shows the street view. This was a typical layout for cinema theatres of all sizes. Alexander Turnbull Library, Wellington, Ref: Eph-A-CINEMA-1914-01

New Zealand musician Harry Shirley has left a vivid account of life in the pit at Auckland's Britannia Theatre during the 1920s:

> At our cinema, the Britannia, there was no way of seeing the film before arranging the music. My solution to the problem was to lay out four or five piles of music on top of the piano, each one suitable for a different mood. There would be gallops for westerns, one steps and foxtrots for comedies, and various pathetic and passionate pieces for dramatic moments. It was quite a trick to read the music at sight, follow the picture and conduct with the right hand. If there was a quick scene change from say, children playing to a baddie lurking round the corner, I had to grab a number from the 'suspense' pile, throw left and right a violin and trumpet part and try to keep some sort of sound coming from the piano until we were ready to start together.[24]

Musicians improvised, cannibalised existing musical scores, and juxtaposed high and low cultural material in ever-changing and creative ways. The use of popular classical pieces, rather than popular songs, was widespread, since the words of popular songs may not have suited the film; purely instrumental music had fewer associations for listeners.

The treatment of classical music in silent cinema often ignored the intentions of its composers or traditional interpretations, and was approached as a way of filling acoustic space appropriately. One anonymous New Zealand film commentator of the 1920s described an American-produced catalogue of classical compositions suitable for films and arranged in 52 categories or moods: 'Grieg was able to tackle most of them, but Mendelssohn was helpful too. Besides Wedding, Funeral Passion, Quietude and National, he covered Aeroplanes. Chopin, it appears, was the man for Monotony.'[25] Hearing classical music at the silent films was in some ways similar to hearing it on many of the contemporary recordings that, owing to their short playing times, entailed the re-arrangement and 'slicing and dicing' of long pieces of music. In this way, the sound and audio technologies of modernity fragmented musical experiences that had traditionally been heard as seamless wholes.

Arranging scores specifically for films was one solution to the problem of finding sounds that matched the rapid scene and mood changes of silent cinema. An arranged score was used for the New Zealand performances of the 1913 religious blockbuster *Quo Vadis*.[26] Based on Henryk Sienkiewicz's 1896 novel, this sword and sandals, lions and Christians epic used a large orchestra as well as a choir in its score. The music had been arranged to suit local conditions by the chief conductor of the Fuller's theatre circuit, E.J. Burke. Publicity for the film stressed that the music had been specially arranged and mentioned that the theatre orchestra had been augmented.[27] A review of *Quo Vadis* mentioned that 'the incidental music adds considerably to the general effect'; little else seems to have been recorded about the augmented orchestra and the choir's efforts.[28]

The Miracle, shown in 1914, also featured a large orchestra and choir. This film version of a 1911 play, written by Karl Vollmöller and directed by Max Rheinhardt, used music that composer Engelbert Humperdinck had written for the original stage version. Many reviewers commented on the film's music, describing it as 'beautiful', 'appropriate' and 'delightful'.[29] Humperdinck was well known in New Zealand for his popular opera *Hansel and Gretel*, and his fame (as well as the film's religious subject) added to *The Miracle*'s credibility. Also in 1914, the music of New Zealand composer Alfred Hill was used in a New Zealand-made film to the same effect. *Hinemoa*, now lost, was based on the legend of Hinemoa and Tutanekai, a story that was to be told several times by New Zealand filmmakers.[30] Hill was famous in both New Zealand and Australia for his cantata *Hinemoa*, as well as for using

Quo Vadis, *an Italian-made biblical blockbuster, featured sensational pictures and carefully crafted music that added to its cultural cachet. At 2.5 hours, it was epic in duration as well as theme and spectacle. Along with 'twenty lions loosed on the Christians', it also featured a chariot race.* Alexander Turnbull Library, Wellington, Ref: Eph-B-CINEMA-1913-01

Māori-derived melodic ideas and motifs in his work.[31] Associating films with the music of well-known composers like Humperdinck and Hill added to their attractiveness and credibility.

By the 1920s most of the movie theatres had 'orchestras' of various sizes that supplied music of all sorts to suit the mood and atmosphere of a particular film. There was no standardised norm of ensemble size or instrumentation. Use of the word 'orchestra' to describe the theatre bands implies a degree of standardisation and grandiosity that does not apply to most of New Zealand's silent cinema musical ensembles.[32]

The orchestra for Whanganui's opera house, before synchronised sound was phased in and musicians were phased out of New Zealand theatres, was a scaled-down version of a full orchestra. The sections such as strings, brass and woodwind were all represented, but by only one or two instruments. The two percussionists may well have supplied sound effects to go with such images as guns firing or people falling over. Many theatre orchestras were even smaller. King's Theatre in Stratford seems to have made do with about half a dozen musicians at most.[33] A piano quartet supplied the music in the Christchurch Grand Theatre in the 1920s.[34] Speedy's Picturedrome at Milford used a small dance band during the later part of the decade to supply music for the films, but only on special occasions. Usually a pianist was employed.[35]

Piano players seem to have borne most of the burden of supplying music for silent films, especially in areas remote from the large cinemas of metropolitan centres.[36] At Birkenhead, then a rural area outside Auckland, films were shown in the Forester's Hall. A local musician, Ted Lanigan, played the piano, and Birkenhead resident Mary Utting recalled that 'when it came to an exciting part he'd play loud and at the soft parts, he'd just play softly'.[37] Alice O'Callahan remembered Lanigan 'never missing a note as he changed from "Melody of Love" to "Napoleon's Last Charge"'.[38] That many other residents of Birkenhead recalled Lanigan's playing suggests the significance of both the movies and the music in their lives.[39]

The most important skill for a pianist playing to a silent film was to match the music with the movie. The pianist p. Hanley, who toured with Montgomery's Pictures and Entertainers in 1905, was praised by one reviewer for 'suitable music for the different series of pictures'.[40] Much hilarity could ensue when images and music came unstuck. Speedy recalled the excitement of the pianist playing the Galop from the 'William Tell Overture' as cowboys headed each other off at the pass. If the pianist missed a mood change, such

These musicians supplied the live soundtracks for the films shown in Whanganui's Opera House during the late 1920s. They were called the Paramount Orchestra (sponsored by Paramount Pictures). The percussion section is quite prominent as sound effects were a major part of live film music. Alexander Turnbull Library, Wellington, Ref: 1/1-017038-F

as a quick cut from pursuit to a love scene, then 'the result could be quite ludicrous', with hysterical laughter all round until the sounds and visuals were realigned.[41] A possibly apocryphal story relates how one New Zealand pianist lost his job after playing 'For He's a Jolly Good Fellow' during the resurrection scene of a film about the life of Christ.[42] Whether by accident or design, musicians might undercut the onstage action for humorous effect, although there is little direct evidence of this being done deliberately in New Zealand.[43]

•••

Some musical practices of the silent film era bridged the gap between musician and audience by including the film viewers in the music. Illustrated songs were one form of such acoustic inclusiveness. These had been popular since the mid-1890s, and involved lantern slides of lyrics and suitable scenes illustrating songs.[44] Illustrated songs were common features of early film shows in New Zealand.[45] These often included audience participation at least during the chorus sections of the music. In 1905 Major Perry's Biorama

Company attracted over a thousand people to Greymouth's opera house with films, the Biorama Band and some singers. The songs included sentimental favourites such as 'The Forger's Daughter' and 'Please Mr Conductor Don't Put Me Off the Train', and featured 'simply beautiful' illustrations with 'the chorus of each song thrown on the screen and the vast audience joining in the singing'.[46] Similarly, the audience at New Plymouth's opera house enjoyed the films of Richardson's Entertainers in 1908. These also featured illustrated songs, 'a form of instructive entertainment that is growing in popularity', and the audience 'heartily joined in the chorus to "Red Wing" and "The Man Who Fights the Fire"'.[47] The slides were made in America and Britain, and this occasionally undermined the effects of the songs when they were performed in New Zealand. One reviewer noted that 'soldier songs of the Goodbye, I'm-going-to-get-shot style are always illustrated in New Zealand with American soldier pictures which is a big mistake'.[48]

A contemporary cartoon plays on the popularity of illustrated songs. 'We Parted on the Shore' was a traditional ballad that Harry Lauder made into a hit by adding to it a large slab of faux-Scottish patter and outrageously rolled rrrrrr's at every chance. Lauder released sheet-music versions, recordings, and even made a short, synchronised-sound Chronophone film of the song.[49] 'We Parted on the Shore' had also been part of the popular 1907 show *Mother Goose*, which used slides to illustrate the song.[50] The cartoon also plays on the fact that the illustrated songs were generally 'familiar, sentimental nostalgic, patriotic'.[51] Parting lovers, orphaned children and family life were favourite subjects. The cartoon reframed and gently satirised this sentimentality with a local scene involving drink and gambling.

Illustrated songs encouraged controlled audience participation and noise. The singers on stage took the lead, with the audience usually joining in the chorus. The words heard were those that were projected on the screen. However, not all audience noises could be controlled so easily. An article by Reynold Ayers of Wanganui that appeared in a New Zealand film magazine in 1921 listed some 'don'ts for movie goers'. Many of these were to do with noise. He enjoined people not to talk too loudly, whistle, stamp their feet when the projector broke down, hiss or jeer villains, eat sweets noisily, shriek with laughter or pass remarks if a child cried in the theatre. Nor should they argue with their neighbour about an actor's abilities and clothes.[52] Ayers' article, while lighthearted, indicated that audience noise could be a problem during silent films. By 1921, most films were being watched in dedicated cinemas, and the protocols that had applied to audience behaviour at

MODERN SONGS ILLUSTRATED: 'We Parted on the Shore', as sung by distressed punters coming from the Takapuna races.

This cartoon played on the popularity of illustrated songs. Two out-of-pocket but happily liquored racing punters sing as they wait for a ferry from Takapuna to Auckland.
New Zealand Observer, 7 December 1907, 5

variety shows in theatres no longer applied. In fact, the 'correct' behaviour suggested in the article was more appropriate to a concert of classical music or a 'serious' play rather than the vaudeville milieu of early film viewing.

It is hard to recover the sounds made by New Zealand film audiences before the talkies.[53] Applause, laughter, whistles and uproar when projection equipment broke down seem to have been just some of the elements involved. Mark Griffen remembered silent films in Foxton as noisy affairs, with heroes cheered and villains 'soundly hooted' to the point where the manager would stop the show, whereupon the 'return to quietness was instantaneous'.[54] Film screenings at Kāwhia included people singing all through the films and abuse aimed at anyone who blocked the screen.[55] Children's matinees were often very noisy from start to finish and Henry Shirley used them mainly as opportunities to practise the piano.[56]

Audience members at screenings of films of the South African War in 1900 often reacted loudly, with 'young patriots distinguishing themselves by their expressions of approval of friends and disapproval of their enemies'.[57] Cheers, boos, applause, catcalls and singing were part of these shows.[58] But when, 16 years later, films of the fighting during the Battle of the Somme were shown in New Zealand, audiences seem to have sat in silence. The British audiences were observed to do the same, although some scenes provoked cheering.[59] Given that World War One lasted much longer than the South

African War, and that New Zealand sustained very high casualties, it may well be that the vaunted realism of the Somme films provoked a respectful silence.

Another factor that might lead to such an assessment is that by 1916 films were mainly seen in dedicated cinemas rather than in variety theatres or halls. A 1915 account described a cinema audience following 'the drama with breathless interest'.[60] By the early 1920s silence may have been expected. However, the need for Reynold Ayers to draw attention to this with a list of unacceptable behaviours, even lightheartedly, indicates that this was not always observed. Whether it was music, sound effects, lecturers or the people around them, New Zealand film-goers during the silent era had much to listen to. The moving pictures were not the entire show. The excitingly modern experience of cinema was incomplete for its audiences without the sounds that sometimes highlighted, sometimes undercut, but always accompanied the flickering images. When the sounds were finally locked in with the pictures as soundtracks, although much was gained many would say much was also lost.

Chapter 12

TALKING PICTURES

•••

Sonic modernism and the technology of modernity collided in Rotorua during the last few months of 1930. The instigator of this mishap was a bespectacled music shop proprietor turned dance musician named Epi Shalfoon.[1] Eric Bierre, a Movietone camera operator, filmed Shalfoon and the Melody Boys as they raced through a honking, clattering, instrumental version of 'He Puru Taitama'.[2] The song, a paean to lust, was often performed by Māori groups as a medium-paced number.[3] Shalfoon and the Melody Boys smeared the song with Dixie stylings, threw caution to the winds and tore it up. After the din subsided, Shalfoon stepped forward and invited viewers to come to hear his band performing 'real' dance music at a venue near them soon.[4] This was probably the first music short made in New Zealand. The canny Shalfoon recognised the power of film as a promotional device. He took advantage of the new and exciting talkies technology to let prospective audience members not only see but also hear what they were in for. The result was the distant ancestor of today's music videos.

> PETER PAN CABARET.
> TUESDAY, MARCH 22.
> GALA EASTER WEEK DANCE.
> EPI SHALFOON and
> HIS MELODY BOYS,
> New Zealand's Crack Dance Combination, As Featured in Fox Movietone News, Whose New Year Carnival Dance Attendance of 1250 Dancers at Rotorua, Constituted a New Zealand Record.
> Open Invitation to All.
> Ladies, 2s; Gents, 2s 6d.
> "Remember, 1250 Dancers Can't Be Wrong."
> W. P. BLACKIE,
> Auckland Representative.

Jazz musician Epi Shalfoon shot New Zealand's first music short entirely as a marketing tool. He made several government publicity films too, which promoted New Zealand as well as his band. Shalfoon grasped the advertising power of the talkies and often mentioned the films in his advertising.
New Zealand Herald, 5 March 1932, 20

Shalfoon made several other film shorts during 1930, but it was not a trouble-free process. He took Bierre to court over the quality of one of these films. According to Eric Roe, Shalfoon's lawyer, the sound was clear enough but the film was so poorly shot that the band's name on the bass drum could not even be read: 'In fact they might have been anybody's boys. Brown's boys or anybody else's boys.'[5] The magistrate agreed. Shalfoon won back the £28 he had paid the camera operator, along with his legal costs.

Shalfoon's musical shorts were not an original idea. The musical short had been developed as early as 1928 by American film companies as a way of marketing songs and artists and so increasing earnings from the companies' interests in recording, broadcasting and publishing.[6] From then and through the 1930s, musical shorts were made by many popular singers and performers such as Jimmy Dorsey, Artie Shaw and Cab Calloway.[7] Many music shorts featured several songs, but at about one minute in duration Shalfoon's were shorter than most. His quick and punchy films made their advertising role quite overt. He and the Melody Boys crossed the frantic and frenetic jazz sound of modernism with the new audio technology of cinematic modernity and fed it back to local audiences through the extremely popular medium of the talkies. Given that there were no commercial recording facilities in New Zealand until 1949, this was a way to get the band's music out to more people than could fit into a dance hall. Shalfoon was cashing in on the popularity of the talking pictures.

The distinctive feature of the talkies was that the heard sounds were synchronised with the pictures. If an image of a violin being played was on the screen, then the audience would hear the sounds of the instrument corresponding to the images of the musician's actions. If a movie actor was speaking, then the words that were heard corresponded to the movements of his lips. However, the introduction of synchronised sound was more of a refinement than a revolution. The historiography of cinema does not treat the years 1928 to 1932, when synchronised sound became widely available, as a watershed or moment of rupture. What is notable about cinema practices before and after the arrival of the talkies is continuity rather than interruption.[8]

But this generalisation is only part of the picture. The arrival of synchronised sound in New Zealand was a complicated process that happened at differing speeds at different places at various times. Some effects, such as the disappearance of live musicians from cinemas, seemed to happen suddenly.[9] But this was not the same experience nationwide.

Orchestras remained important attractions for some cinemas after the arrival of the talkies.[10] And some New Zealanders objected to the talkies on both moral and aesthetic grounds, and were therefore reluctant to embrace the new medium.[11] The sounds of the talkies echoed through New Zealand's culture as well as the country's cinemas.

One theme that clearly emerges from all this is standardisation. The sounds on a talkie's soundtrack never varied, no matter where the film was screened. If the soundtrack featured a symphony orchestra, then an orchestra was heard even if the film was being shown in a small town hall where the only musical instrument available for silent films had been an upright piano. Overall, the talkies 'canned' cinematic sound in the same ways in which records 'canned' sound. Like records too, a cinema soundtrack was durable, portable and repeatable. The technologies of acoustic modernity (records, radio, films) standardised sonic experiences at different times during the first half of the twentieth century. This happened to films during the late 1920s and early 1930s when dependable, synchronised sound systems were developed and deployed throughout the world by large American entertainment corporations. But the reactions to 'canned' cinematic sound in New Zealand were at times far from standardised. New Zealand listeners at the talkies may have all heard the same sounds at the country's cinemas, but they often listened to these sounds in different ways.

• • •

Going to the movies has been a popular pastime in New Zealand ever since the technology of the talkies arrived in the 1930s.[12] Official figures for attendances 1939–40 indicate that each person in the country averaged 19 movie admissions per year.[13] Significantly, most were now hearing the same words and music that were heard all around the world. And these were mainly American. Between 1929 and 1940, 1883 films were released in New Zealand, 81 per cent of which were made in the United States.[14] The musicals, dramas, comedies, westerns and romances mass produced by Hollywood studios dominated New Zealand's movie screens.

The American companies which owned the new sound synchronisation technologies kept the machinery secret to maintain control over the production of talkies. As a result, the first New Zealand sound movie efforts were literally homemade. Edwin Coubray, a New Zealand cameraman, designed and built his own sound-on-film system and successfully trialled it in 1930.[15] This system eventually came into the hands of Jack Welsh, a

Dunedin photographer and film maker. Welsh used the equipment to make three feature-length talkies, none of which received much exposure outside Otago and Southland, and by 1937 was using the system to make news reels.[16]

The domination of New Zealand's theatres by American films made it difficult for any local productions to compete. The American studios had seemingly unlimited financial, human and technical resources, and produced slick and spectacular films. A typical Hollywood studio used multiple cameras and takes to get a single shot. Hollywood stars such as Clark Gable, Shirley Temple, Fred Astaire, Ginger Rogers, Jean Harlow and Errol Flynn were sure-fire attractions for audiences around the world.

By contrast, when the New Zealand director Welsh made his first feature film, *Down on the Farm* (1935), he had one camera, and limited shots to two takes to save money. With no studio lighting he had to film outside, and was dependent on good weather and hours of sunlight. The sound system itself had a limited range, so actors had to group together near a microphone. They were all from Dunedin theatrical circles and none had any experience in film.[17] The sound system also picked up wind noise, exacerbating the difficulties of outside shooting. While the film achieved some local success, this may have been due to the novelty of seeing local people and places on the silver screen. Outside of Otago, the film flopped.[18]

Another early New Zealand sound film maker was Rudall Hayward. Hayward had a long and varied career that extended from his first silent film in 1920 to his last feature in 1972. His work ranged from short documentaries to full-length feature films and has received much scholarly attention.[19] Hayward, like Coubray, built his own sound-on-film equipment and first used it in *Hamilton Talks* (1934), a community comedy.

Hayward produced 23 community comedies between 1928 and 1930. These silent two-reel films were made very quickly and used stock plots, characters and intertitles. Hilda Hayward, Rudall's first wife, would arrive in a town and quickly select lead characters after auditioning various local worthies and personalities.[20] Crowd shots were used as often as possible in an effort to get a large number of people into the film. Popular locations were chosen for ready-to-hand sets. The turnaround on these films from start to screening was a week to 10 days. The result was a local sensation and guaranteed good audiences, at least in that town. By the time the interest had faded, the Haywards would be on to the next town.[21]

New Zealand's pioneer film maker Rudall Hayward used this sort of advertising to get local personalities to appear in his 'local community films'. Apart from the excitement of being in the movies, the leaflet held out the promise of future stardom for those 'showing any promise'. Ngā Taonga Sound & Vision, D0516 13/1

The soundtrack for *Hamilton Talks* uses music between scenes and over some establishing shots. The ambience for this differs markedly from the scenes involving actors. The intensity of the sounds varies and some spoken words are indistinct. The overall sound quality of the film is quite low compared to most commercially made films of 1934. But the attraction of the film to its audience was that the voices were those of local people. Hearing and seeing themselves on the screen was a novel experience for most New Zealanders during the 1930s. Or at least seeing and hearing some of themselves: Māori are absent from all of Hayward's community comedies, which portray a white New Zealand enjoying such fruits of modernity as cars, aeroplanes and modern houses and buildings. Māori were not part of this vision of the country shared by most Pākehā during the 1930s.[22]

Hayward's feature films such as *My Lady of the Cave* (1922), *Rewi's Last Stand/The Stand* (1925/1940) and *The Te Kooti Trail* (1927), by contrast, explore the relations between Māori and Pākehā, and deal with the complex process of colonisation as historical and ongoing. The voices and sounds of Māori and Pākehā are equally important in the talkie version of *Rewi's Last Stand*, made in 1940.

The 1920s saw many attempts to develop a reliable technology that synchronised sounds and images.[23] The system that eventually came to dominate talkie theatres was made by the American company, Western Electric.[24] It initially came in two forms. The first was a sound-on-disc technology which involved playing records that were synchronised with the film reel. It was difficult to keep the images and sounds in time, as the records had to be precisely cued. The records were also delicate and became harder to synchronise as they became worn.[25] Napier's Gaiety Theatre used Western Electric's sound-on-disc system and 'audiences often burst into laughter when bad picture-sound synchronisation projected male voices over female characters'.[26] Technology malfunctions might provide unintentional laughs for some, but they undermined the illusion of realism that film depended on for its full effect.

Western Electric's other system was sound-on-film. This technology involved recording sound onto the film rather than utilising a sound recording that was synchronised to play with the films. The sound-on-film system provided sound without the hiss of the record's surface, and also kept the sounds synchronised with the picture images.[27] The Western Electric sound-on-film system faced no real competition or opposition in New Zealand and spread rapidly through the country's cinemas, although it was expensive to install and run. The installation might cost between £2000 and £5000 according to the venue's size, and operators also had to pay a weekly rental of £5 to £7.[28]

The talkies revived the movie business and led to a spate of theatre building and conversion which created an over-supply of venues around the country. Combined with the effects of the Depression, this movie investment boom in New Zealand was followed by a bust during the early 1930s.[29] But plenty of theatres survived and more were built so that by 1939 there were 576 for a population of 1.6 million people.[30] New Zealanders were flocking to the movies, whether in brand-new purpose-built theatres or older venues that had been converted into cinemas.

The arrival of reliable synchronised sound systems for motion pictures required older venues to install new equipment that could project the pictures and play the sounds of the films. Projection rooms had to be soundproofed, projectors fitted with sound-heads or replaced, and amplifiers and speakers installed in the venue.[31] This in turn meant that the structure of the film venue might have to be altered to suit the new acoustic requirements.[32] The owners of the Picturedrome cinema at Milford resorted to 'lowering the ceiling with a sound absorbing material' and draping crepe de chine over the walls, a process that was 'risky and expensive'.[33] But this method worked and the theatre was advertising its 'perfect sound, new decorations' in June 1936 when it reopened with *Broadway Melody of 1936*.[34] The small Taranaki town of Eltham used the local town hall as a cinema. With the advent of the talkies, two new projectors were installed and the entire hall rewired to supply reliable direct current for the equipment. Two record turntables were installed along with a large amplifier.[35] The changes were appreciated by Eltham's film-goers. Eva Hargreaves recalled that the town's main street was lined with cars 'when there was a big movie on' and described the talkies as 'wonderful, really beautiful'.[36] Seeing and hearing movies in a rural town hall was just as pleasurable and offered every bit as much of a sense of occasion as a big city theatre.

The cinemas that were built as the talkies arrived were designed for the acoustic and technological requirements of the new type of movie. The advertising for the opening of Auckland's Civic Theatre on 20 December 1929 talked up the acoustics and 'crystal-clear dialogue and music' delivered by the venue's Western Electric sound system, the 'finest talking picture plant it is possible for science to produce'.[37] The contrast between Eltham's town hall and the gilded splendours of the Civic could not have been greater. The Civic seated 3500 patrons under a ceiling that displayed twinkling stars and drifting clouds. The decor was a lavish mixture of Hindu and Moorish styles, complete with statues of elephants, tigers, Buddha and a pair of blue-eyed panthers on either side of the stage.[38] One reporter called it a 'veritable enchanted palace' and likened it to 'entering an Eastern potentate's palace in a country where roofs were unknown'.[39] But audiences heard the same sounds from a talkie whether it was played in the Eltham Town Hall with its homely wallpapered walls, or Auckland's Civic with its twinkling stars and fake clouds scudding overhead.

Hearing these 'crystal-clear' sounds delivered by the latest recording technology was part of the attraction of the talkies. The traditional experience of going to the pictures was transformed. To experience the talkies was to experience the most up-to-date expression of technological modernity, delivering recorded sounds that were free of the rumble and hiss of gramophone records. The Western Electric system was the outer limit of sound technology, at least according to the Civic's publicity.

Auckland Civic Theatre's opening programme was a mixture of films, ballet, orchestral selections and a recital on the newly installed Wurlitzer organ.[40] The main talkie was *The Three Live Ghosts*, a British-made comedy. This was no accident: the Civic's management had carefully chosen it as being 'essentially English as opposed to American' to appropriately recognise 'New Zealand being British to the backbone'.[41] British or not, the film does not seem to have made much of an impact. One review of the Civic's opening dwelt at length on the decor, the orchestra, the dancers and the organ but managed only a brief description of the main film in the final paragraph.[42]

In fact, the real business was happening across Queen Street at the St James Theatre, where the Warner Brothers-produced film *Gold Diggers of Broadway* was packing in the crowds. The St James converted to sound during December 1929 and reopened on the 26th, just three days ahead of the Civic. *Gold Diggers of Broadway* was a landmark in the development of the American movie musical. The witty script balanced romance and jokes, the strong cast delivered a varied set of musical numbers, and the Technicolor was clear and sharp.[43] It was a huge hit in New Zealand. No reliable box-office statistics survive, but by the end of January 1930 the St James was advertising that 122,977 of Auckland's 230,000 people had seen *Gold Diggers of Broadway*.[44] By the last two nights of the run the figure was 179,032.[45] American glamour seemed very attractive to New Zealand audiences in 1930. They took pleasure in the film's stylish costumes, sets and dances, but also engaged with the soundtrack in a variety of creative ways that were both commercial and personal.

The popular songs 'Tip-Toe Through the Tulips' and 'Painting the Clouds With Sunshine' were performed by Nick Lucas, who played a guitar as he sang. Walter Smith, an Auckland music teacher, tried to cash in on the film by pointing out that *Gold Diggers of Broadway* was full of guitars and that he could 'teach you to play the same way'.[46] Smith was not alone in piggy-backing on the popularity of the music heard in the film. Nick Lucas recorded for Brunswick Records, a label that Warner Brothers bought during 1930. Rival recording companies quickly produced cover versions to cash in on the songs' success.

In its advertisements for these, the Regal label offered songs from a

Wellington department store Bannatyne and Hunter offered record buyers a wide selection of hits from the musical talkies that played in the city during 1929. These included Sunny Side Up, On with the Show *and* Broadway Melody, *which was the first talkie to feature Technicolor, and also the first to win an Academy Award.* NZ Truth, *20 February 1930, 18*

number of musical films performed by various artists. Only a few of the artists were named and it seems that the songs, rather than the performers, were the main selling point. Most had featured in film musicals that played in Wellington during 1929. Recordings of these songs could be played at will, wherever desired, and might work as a memento of the film. The Columbia advertisement appeared during the Wellington run of *Gold Diggers of Broadway* and used the film as a selling point, although none of the performers on these records appeared in the movie.

Not everyone was impressed by the talkies' combination of synchronised sound and image. Some New Zealanders believed that moving pictures were diminished by the added sounds. Some objected to the particular sounds that went with the pictures. Still others heard in the sounds of the talkies a menace to the nation's moral character, especially among young people. Concerns were expressed that the talkies were manufacturing thousands of flappers 'of the naughty disposition'.[47] The films were 'a moral scourge to the community'.[48] George Spooner, chairman of the Auckland Primary School

Record labels like Columbia and HMV quickly released records that cashed in on the all-singing, all-talking, all-spectacular Gold Diggers of Broadway. *Buying both records would have meant getting two versions of the biggest hit from the show, 'Tip-Toe thru' the Tulips'. The talkies thrilled a lot of record executives as well as filmgoers.* Evening Post, 24 April 1930, 5

Committees' Association in 1930, warned that talkies were being shown 'that to the young mind, gave them a disregard for home life and the sanctity of things for which we, as a British community, stood'.[49]

These concerns shared a common desire to preserve a familiar idea of 'Britishness' against an imagined idea of America based on the more sensational aspects of mass culture that were exported to the rest of the world. Historian James Belich interprets this rejection of American popular culture as part of a process of recolonisation that saw New Zealand renew its cultural ties with Britain. Middle-class and official cultural gatekeepers were engaged in an ongoing struggle to exclude American mass-culture products from New Zealand and to control those that slipped through the filters of censorship. These products included American cars, radio serials, novels, comic books, records and films.[50] The popularity of Hollywood films in New Zealand went against the idea of 'British at all times', and provoked what historian Simon Sigley described as an 'almost hysterical reaction'.[51]

The objections to American films, especially the talkies, during the early 1930s took place against a social backdrop of xenophobia and conformity.[52] A quota system, under which 20 per cent of the films imported into New Zealand were required to be British, was introduced by the New Zealand government in 1928.[53] New Zealand theatre owners willingly did their patriotic duty by buying British films, but ran into supply problems: the British film industry could not supply the numbers of films needed to keep their theatres turning over fresh attractions.[54] Another problem was audience demand. American films were regarded as being better made and generally more exciting and glamorous than British productions. Some moviegoers would simply skip the opening half of a film programme if it contained British films.[55] American films thus continued to dominate New Zealand's cinemas despite official disapproval.

Given the linkages between film marketing and record and sheet music sales, it may well be the case that a large proportion of the music heard in New Zealand homes was American. But this is difficult to prove as there were no record charts in existence.[56] However, in terms of popular culture, the links with Britain postulated by the recolonisation thesis seem to have been eroding rather than strengthening during the late 1920s and the early 1930s when synchronised sound was introduced to New Zealand.

New Zealand's defenders of high culture looked down on films in general but particularly looked askance at American films. The student magazine for Wellington's Victoria College ran the following acrostic poem shortly after the introduction of the talkies:

> A are the automobiles we import,
> M are the movies where children are taught
> Endless inanities, hogwash and crime,
> R is the racket of jazz and rag-time,
> I is the influence increasing each day
> Changing New Zealand to New U.S.A.,
> A are the authors like Zane Grey the great,
> N are the novels that pay for his bait;
> I, the ideas that the talkies impart,
> S is their standard in music and art;
> E, the example is civilization
> Displayed by the cultured American nation.[57]

This sort of low opinion of the talkies was registered by the New Zealand-born historian J.C. Beaglehole, who saw his first talkie while completing his doctorate in London in 1929. He wrote about the experience to his parents in no uncertain terms. The film was 'appalling in construction, speech, song and everything else'. He had forgotten that 'Yank pictures' were 'shockingly bad beside the European ones'.[58] The comparison between the American film and the 'European ones' seems to have been based on the idea that 'Yank pictures' typically had nothing to offer a serious viewer, and were focused instead on a cavalcade of crime, violence and sex that rotted young people's minds and threatened their morals.

Some of New Zealand's serious viewers were not so quick to condemn 'Yank pictures'. John Storm of the Auckland Film Society (AFS), praised *Gold Diggers of Broadway* for its acting and humour, though added a satirical twist by calling it 'a sort of Guy Fawkes night out for children over thirty'.[59]

The 23 articles written by members of the AFS that appeared in the *Auckland Star* approached American films in a fairly open-minded manner.[60] But they also reflected critical ideas about cinema just as synchronised sound became the medium's norm, and voiced common fears about how sound might affect the art of film.[61] One of these was that sound would eliminate the expressive features that made cinema unique and would reduce the medium to the 'imitation of stage plays'.[62] By voicing such concerns, these New Zealanders were echoing the criticisms of sound film made by European writers like the German media theorist, Rudolf Arnheim.[63]

As synchronised sound was deployed in film studios and film theatres around the world, the argument advanced by Arnheim and other European film theorists was that mimetic sounds, imitating reality, would reduce cinema to a flat imitation of theatre.[64] Some New Zealand film viewers, such as the AFS writers, shared these concerns but generally greeted synchronised sound as a promising development for cinema rather than as a threat to the British tone of New Zealand's culture.

While the cultural and moral dangers of the cinema had been vigorously debated by New Zealanders for many years, the arrival of the talkies raised new concerns about the effects on New Zealand's spoken language, and especially young people's language.[65] American accents and slang were heard as a threat to the British language that was New Zealand's ideal.[66] Writing under the name of 'Pure English', a concerned citizen from Wellington wrote to the editor of the *Evening Post*, objecting to the actors' accents, pronunciation, and unmodulated and raucous voices like 'the sound a rip-

Father: Tommy, I hear you've been smoking!

Tommy (a student of gangster talkies): Say, dad, what's the low-down ... has ma squealed?
—"Tatler."

Some concerned parents and other authority figures feared the worst for the English language under the cacophony of Yankee slang that had begun to sound through the nation's cinemas. Auckland Star, 28 November 1931, Supplement, 2

saw would make going through a petrol can'.[67]

Yet such views were far from universal or uncontested. Several replies to 'Pure English' made a case for the talkies. 'Celluloid' pointed out that the American accent no more murdered English than Cockney. New Zealanders mispronounced vowels and had their own slang: 'too right', for example, was as idiotic as anything from America.[68] A letter from 'New Zealand Born' accused 'Pure English' of being envious of the better quality of American films. This writer also praised the clarity of talkie actors' speech, and scoffed at the idea that the accents heard were too broad.[69]

These letters are a microcosm of the sorts of reactions the talkies aroused. 'Pure English', with his or her carefully chosen pen name, heard the sounds of the films as an alien incursion that threatened the homogeneity of New Zealand's British-based culture. This writer viewed pure English language as the key to pure English culture. But the writers using the names 'Celluloid' and 'New Zealand Born' were more relaxed about the medium and regarded talkies as harmless and pleasurable diversions. 'Pure English' saw audience members as passive and easily affected by the talking pictures. 'Celluloid' and 'New Zealand Born' wrote as active listeners who took what they wanted from the films.

The discussion about the stealthy Americanisation of New Zealand by the talkies went beyond the pages of newspapers all the way to the highest forum of political debate. The matter was raised in the Legislative Council by the scientist and educationalist George Thomson, who favoured censoring the talkies and raising the quota percentage of British films. Thomson was speaking on behalf of concerned teachers and bodies such as the National Council of Women. But the council members were unimpressed, noting that gramophone records were not affecting speech patterns and that the talkies were a new invention and would improve in time.[70] The quota for British films was one thing; it was quite another to prevent New Zealanders from seeing American films. Objectionable material could be dealt with through censorship.

Censoring the sounds of the talkies had been built into the 1928 Cinematograph Films Act. This Act allowed the censor to 'take such reproduction of sound into consideration in determining whether or not the film should be approved, or in determining what excisions, if any, should be made therefrom'.[71] It gave the censor powers to remove sounds that might be offensive, whether spoken, or musical or sound effects. But it was not the sounds alone that were cause for concern. Along with colour, sounds made the films more realistic, adding to the potential for the film to become dangerous or offensive. A 1930 report for the Department of Internal Affairs, the body that handled film censorship, stated that 'the more vivid presentation of life made possible by the addition of sound and colour has made it necessary to issue an increased number of certificates recommending films as being more suitable for adult audiences'.[72] The issue of the effects of American culture on New Zealand was not a factor that influenced the film censor. The 'vile American accent' continued to be enjoyed by New Zealand film-goers.[73]

New Zealand film critic and reviewer Gordon Mirams wrote in 1945 that Anzac, Hollywood and Home (meaning Britain) were inscribed on New Zealand's national heart, and it would be rash to say which was most important.[74] Mirams' point was that America's cultural products, especially films, were greatly admired by New Zealand audiences, and did not diminish New Zealanders' loyalty to country and king. Cultural nationalism was unimportant for many film-goers: they went to see American films because these films were more exciting, more thrilling, funnier and sexier than those made in Britain. The language, slang, jokes and songs on the soundtracks were the sounds of an imaginary America that existed only on the screen. For those who enjoyed this fantasyland of gangsters, cowboys and showgirls, using these words and jokes with friends or playing a record of a song from a romantic comedy was an addition to New Zealand's culture rather than a subtraction. Similarly, historicising cinema in New Zealand through a nationalist viewpoint that concentrates only on locally made films, and those who made them, ignores the cinema that most New Zealanders valued and enjoyed in the past. Most of the history of New Zealand's culture of film-going has an American accent.

•••

The New Zealand accent was hardly heard from cinema's speakers during the early 1930s when the talkies became widespread. Most of New Zealand's film making during this decade was of tourist or technical features and news reels. A few feature-length films were made, but New Zealand's sounds were drowned out in the country's cinemas by the films of America and Britain. Local film makers usually had to make do with few resources and no guarantees of audiences.

The talkies were quickly taken up by New Zealand's cinema industry and audiences. When the National Union of Students had its annual ball in 1936 (at which many arrived late, having first gone to the pictures), there was, as always, a novel twist to the proceedings. A room was turned into a movie theatre and silent films were shown, including cartoons of Mickey Mouse and Felix the Cat as well as a comedy by Harold Lloyd.[75] By 1936 watching a silent movie had become an amusing novelty. That night in Wellington, lovers of dance and song might have seen *Follow the Fleet* with Ginger Rogers and Fred Astaire at Our Theatre in Newtown. Those with more macabre tastes could have watched Boris Karloff in *The Walking Dead* ('Risen from the dead to avenge his own murder') at the Rex Theatre on Cuba Street. A romantic comedy was on offer at the Paramount, where Clark Gable and Claudette Colbert starred in 'the brightest show in town', *It Happened One Night*.[76] The theatre's Wurlitzer organ supplied music between shows.

Behind the standardised experiences of canned sounds and images there was, however, turmoil as New Zealanders assimilated the talkies into their visual and auditory worlds. The talking pictures called forth new responses to the experience of cinema, an important part of New Zealander's leisure. Some heard the talkies as a threat; others heard them as novel and exciting entertainments. There were varied responses to the aesthetics of the new cinematic technology.

The sounds of the talkies required new projectors, amplifiers and speakers, which had major effects on theatre design and layout. Existing theatres were modified and there was a spate of new theatre-building. The recorded sounds of the talkies also put many musicians out of a job and bolstered record sales, as entertainment companies cross-marketed sheet music and 78s through musicals and film shorts.

In hindsight, the adaptation of the talkies looks inevitable and relatively straightforward. We are so used to synchronised sound and vision, it seems odd that anyone would have questioned the technology. However, this view of history leaves out the cultural static, distortion and noise that go with the arrival of any form of audio technology, including the talkies.

Conclusion

THE DIN OF THE WORLD

∙∙∙

Maud Taylor was born in London in 1879 and her family emigrated to New Plymouth in 1891. She was a talented singer, and performed in choirs and as a soloist. She worked as a school teacher and married Frederick Basham, a civil engineer, in 1904. The couple had three children and eventually settled in Ngatea on the Hauraki Plains. Maud died in 1963. This sounds like the underpinning of an unexceptional life, but Maud was a major media star. Better known as Aunt Daisy, from the late 1920s until her death she made radio, wrote books and newspaper columns, appeared in news reels, and was part of early television broadcasts in New Zealand. Her distinctive rapid-fire voice and her cheerful 'Good morning everybody' catchphrase were part of daily life for hundreds of thousands of New Zealanders, and she remains a nostalgic symbol in popular culture.

Maud Basham's involvement with television heralded the arrival of a new medium that in many ways supplanted radio, taking over such content as soap operas, drama, news, documentaries, sport coverage and music, and adding pictures. Many who worked in radio, from hosts such as Selwyn Toogood through to technicians and administrative staff, also found niches in the developing television sector in New Zealand. Gradually television became the domestic electronic hearth and centre of evening entertainment. Meantime, new listening patterns developed as radios became smaller and cheaper throughout the post-war years. Radio was heard in cars, in bedrooms, on beaches and in other locations. Television grew to dominate the evening hours and radio became important in the mornings.

But one medium did not supplant the other. Radio and television in New Zealand had (and have) complex linkages, interactions and uses. The sounds of television were added to those from radios and records in the home, and New Zealanders used them in sometimes surprising ways. I remember my brother watching cricket on television with the sound down while listening to the radio broadcast of the game. The radio commentary was more detailed and informative than the televised commentary, as the pictures were felt to be sufficient for viewers. I often think of television as radio slowed down by pictures.

Television itself is now undergoing major changes as digital technologies transform media consumption habits globally. Social media, tablets, phones and other technologies allow new patterns of media consumption, and are profoundly altering how we think about media and the way we use them in our lives. But these technologies all use recorded sounds to inform, educate and entertain. They have, too, been miniaturised and their capabilities vastly expanded, just as other technologies have since the late nineteenth century. P.J. Bohanna set up the New Zealand record industry in 1900 with 5000 recordings. Nowadays these would all happily fit on a modern phone with storage to spare, and a streaming account will make more music available than could be heard in a lifetime. These sounds can be heard anywhere, played over and over, and will be there as long as the phone works or we pay our streaming subscription fees. The sounds are portable, repeatable and durable in ways they were not before the phonograph.

But digital technologies are not necessarily changing the fundamental characteristics of hearing recorded sound that were developed during the period covered by this book. We are watching and hearing more varied images and sounds, and in greater quantities, but the qualitative characteristics of listening to technologically mediated audio are not a great deal different. A phone crammed full of mp3s is a 'phonograph' in the sense that the word means to 'write sound'. The sounds are captured, stored, preserved, and can heard over and over in any place. The revolution in sound is not that we have so much of it available but that we can 'write' it and hear it anytime and anywhere as often as we want. This is what made the new audio technologies so important and exciting for their listeners.

New Zealanders were part of this revolution. They engaged with the sounds of the world and made their own contributions to the global din of records, radios and films. Taking all sorts of new sounds from elsewhere, they made local versions that went beyond copying and became important parts of the local soundscape. The famous 'Maori Battalion Marching Song' of World War Two was based on the melody of a US college football song, 'The Washington and Lee Swing'. With new lyrics and some musical tinkering it became a powerful sonic signifier of New Zealand and Māori culture in particular. Global sounds were hybridised and retooled for local purposes and meanings. Songs heard in musicals became part of New Zealand's popular culture through records, radios and the films they featured in. Many New Zealanders heard the world's din as entertaining and enriching.

For some, the new audio technologies threatened the moral fabric of society and the physical health of their listeners. Just as many concerns are expressed about the effects of modern digital technologies, so were the gramophone, wireless and cinema considered by some as detrimental to the health of individuals and society. The 'mechanised pleasures' of the modern world dulled and impoverished their auditors, according to Elizabeth Connors in the *Mirror* in 1931:

> Turn a button and start the radio, wind a handle and listen to the gramophone, sit in your seat and watch mechanical pictures ... after a few years of this form of entertainment, absolute boredom sets in. This is only too apparent in the discontented faces one sees all around. Pleasure has become too cheap, too easy of attainment to afford the appreciative interest it should.[1]

Such jeremiads express complicated sets of ideas and fears about technology, culture and society. The composer, philosopher and acoustic environmentalist Murray Schafer argues that the separation of sound from source is an alienating and dehumanising side-effect of recorded audio. He hears audio technologies as unwanted intruders into natural soundscapes. Schafer coined the term 'schizophonia' to denote the separation of sounds from their sources.[2] For him, schizophonia is a malaise of sonic modernity that has unfortunate consequences for individuals and societies.

Schafer's view of a 'natural' world of sound as opposed to an artificial and polluting world of mechanically produced sound is built on the idea that sounds and bodies are ideally united in a holistic relation. In this view, the self is a united presence that has been fragmented by the experience of technological modernity. Many see digital technologies as having similar effects on modern lives and cultures. However, other scholars of sound such as Jonathan Sterne argue that this view removes the possibility of historicising the subject, and imposes an artificial idea of the subject and the body. Schafer's view makes audio technologies invisible, in that he sees the source of a sound as outside the sound-reproducing technology. The sounds are then heard as copies of an imagined source and, as Sterne phrases it, 'technology vanishes'.[3]

This book has brought the technology back into view, and heard listening through records, radios and films as embedded in social practices and discursive formations. Recorded sound was not alienating or dehumanising for most New Zealand listeners. The sounds were creatively incorporated into a set of rich and varied leisure and educational practices. For many individuals, recorded sound enhanced their participation in a protean and exciting international popular culture. Listening was a way of engaging with the world and making new sounds that combined the foreign with the local to produce new experiences and sensibilities.

Rediscovering the listening practices associated with the gramophone, radio and cinema provides a genealogy for the ways in which digital devices are heard. The earphone is a ubiquitous piece of technology that is emblematic of digital culture. Many discussions about portable sound concentrate on devices that use earphones such as the Walkman and the iPod, and point to the ways in which personal and public spaces are mediated through portable music players, often to offset feelings of alienation and isolation.[4] The machines create private aesthetic spaces, literally within the listener's head.[5] Such accounts stress the novelty of the sort of interior spaces created by earphones, but this ignores the fact that many early audio technologies also required the use of earphones, and might involve several listeners at once. Their listeners may well have experienced the same kind of inner musical spaces that are associated with modern earphone listening. 'Modern' listening experiences have surprisingly long, complex and varied genealogies that are not purely dependent on reaching some abstract and absolute level of technological progress.

New Zealanders literally heard the world through their gramophones and radio sets, and at their local cinemas. They were willing participants in an international mass community of technologically mediated audio consumption and leisure. The scratchy blare of sonic technologies was the sound of global modernity, and we are still living through its consequences.

Modernity defies a single definition. It is a complex concept that refers to a matrix of trends and events that some authors date back to the start of the sixteenth century.[6] However, some commonly accepted characteristics include mass leisure and consumption, rationality, industrialisation, urbanisation, the nation state, democracy, mass media, faith in science, technology and progress, widespread education, the rise of global commercial entities, and a general sense of ceaseless and dynamic change.[7] Ideas and experiences of time and space changed radically, too, as transport and communications technologies seem to have made the world smaller and life faster.[8] The concept of modernity embraces all of these ideas, along with new sensations of uncertainty and contingency.

The instability aroused by the audio experiences of modernity was captured by the novelist J.B. Priestley after he first heard a ragtime band in a Leeds music hall in 1913: 'We were yanked into our own age … the end of confidence and any feeling of security. Here was something new, strange, curiously disturbing, chanting and drumming us into another kind of life in which anything might happen.'[9] Priestley was looking back after 50 years, but his account captured the shockwaves of the sounds of modernity that, in themselves, echoed many of its other characteristics. Industrial products marketed worldwide by global entertainment companies, the sounds and the new sonic machines were tokens as well as symbols of scientific and technological progress. The technologies of audio modernity transformed education, consumption and leisure through new modes of listening.

New Zealanders' listening practices and cultures were also transformed by this global experience of audio modernity. Listening became fragmented and repeatable. Records could be played over and over. Listeners might listen to one section of a symphony followed by a jazz tune followed by a laughing song, and these might be repeated in any order. Radio brought into houses a variety of sounds that ranged from classical music to dance tunes. Listeners could tune into anything from a New Zealand station to a shortwave broadcast from the other side of the world. Musicians working in local cinemas cut up music to supply appropriate sounds for whatever was on the screen. This practice continued with the advent of cinema soundtracks during the late 1920s, and the sounds could be heard again on repeated visits. Popular film songs and music were also available on record. Shuffling sounds was an accepted part of listening long before the iPod.

Before recording, the experience of musical sound was linear. Musicians began at the start of a piece and, all going well, played through to the end. Listeners would hear musical works as singular temporal entities in a single space. But audio technologies challenged this order. The portability, durability and repeatability of sound made possible by the new technologies allowed for partial, contingent and fragmented modes of listening in any number of spaces. According to sonic historian Jonathan Sterne, 'Modernity marks a new plasticity in the social organisation, formation and movement of sound.'[10] This plasticity was part of the experience of New Zealanders as they played records, tuned radios, and sat in cinemas. The modern world was fast, noisy and disjointed. And so was listening.

Listening is a private experience. It is a mental experience that arises when kinetic energy is transformed into electrical impulses by our ears and these currents in turn are used by our brains to create the experience of sound in our consciousness in ways we do not yet fully understand. Sound is all around and through us as vibrations, but listening is within us. As the film director Robert Bresson wrote, 'The ear goes more toward the within, the eye toward the outer.'[11] Writing history about such a private, fleeting and seemingly ephemeral experience as listening may, at first, seem an impossible task. But the past was noisy and still reverberates through the ways we live now. We just need to listen.

NOTES

Preface

1. *New Zealand Herald* (*NZH*), 28 November 2012, www.nzherald.co.nz/entertainment/news/article.cfm?c_id=1501119&objectid=10850438
2. Alejandro Zentner, 'Online Sales, Internet Use, File Sharing, and the Decline of Retail Music Specialty Stores', *Information Economics and Policy*, 20, 3, 2008, pp. 288–300.
3. Nicky Harrop, 'Marbecks', 24 February 2015, www.audioculture.co.nz/scenes/marbecks
4. Richard Osborne, *Vinyl: A history of the analogue record*, pp. 1–2.
5. Emily Chivers Yochim and Megan Biddinger, '"It kind of gives you that vintage feel": Vinyl records and the trope of death', *Media Culture & Society*, 30, 2, 2008, pp. 183–95; Osborne, *Vinyl: A history of the analogue record*; Tomi Nokelainen and Ozgur Dedehayir, 'Technological Adoption and Use After Mass Market Displacement: The case of the LP record', *Technovation*, 36–37, 2015, pp. 65–76; Dominik Bartmanski and Ian Woodward, *Vinyl: The analogue record in the digital age*; David Sarpong, Shi Dong and Gloria Appiah, '"Vinyl never say die": The e-incarnation, adoption and diffusion of retro-technologies', *Technological Forecasting & Social Change*, 103, 2016, pp. 109–18.
6. Record Store Day, www.recordstoreday.com/
7. James Lastra, *Sound Technology and the American Cinema: Perception, representation, modernity*, p. 4.
8. The intersections between noise, music and culture have been studied from a variety of perspectives. See, for example Raymond W. Smilor, 'Toward an Environmental Perspective: The anti-noise campaign 1893–1932', in Martin V. Melosi, ed., *Pollution and Reform in American Cities 1879–1930*, pp. 135–51; Lawrence Baron, 'Noise and Degeneration: Theodor Lessing's crusade for quiet', *Journal of Contemporary History*, 17, 1, 1982, pp. 165–78; Jacques Attali, *Noise: The political economy of music*; Peter Bailey, 'Breaking the Sound Barrier: A historian listens to noise', *Body & Society*, 2, 2, 1996, pp. 49–66; Stan Link, 'The Work of Reproduction in the Mechanical Aging of an Art: Listening to noise', *Computer Music Journal*, 25, 1, 2001, pp. 34–47; Karin Bijsterveld, 'The Diabolical Symphony of the Mechanical Age: Technology and symbolism of sound in European and North American noise abatement campaigns 1900–40', *Social Studies of Science*, 31, 1, 2001, pp. 37–70; Thompson, *The Soundscape of Modernity*, pp. 115–68; Bijsterveld, 'The City of Din: Decibels, noise and neighbours in the Netherlands 1910–1980', *Osiris*, 2nd series, 18, 2003, pp. 173–93; John Picker, *Victorian Soundscapes*, pp. 41–81; Nick Smith, 'The Splinter in Your Ear: Noise as the semblance of critique', *Culture, Theory & Critique*, 46, 1, 2005, pp. 43–59; Beth Meszaros, 'Infernal Sound Cues: Aural geographies and the politics of noise', *Modern Drama*, 48, 1, 2005, pp. 118–31; Paul Hegarty,

Noise/Music: A history; Nick Yablon, 'Echoes of the City: Spacing sound, sounding space, 1888-1916', *American Literary History*, 19, 3, 2007, pp. 629-60; Karin Bijsterveld, *Mechanical Sound Technology, Culture and Public Problems of Noise in the Twentieth Century*; Hillel Schwartz, *Making Noise: From Babel to the big bang & beyond*.

9. See, for example, Steven Connor, 'The Modern Auditory I', in Roy Porter, ed., *Rewriting the Self: Histories from the Renaissance to the present*, pp. 203-23; Jennifer Forest, 'Scripting the Female Voice: The phonograph, the cinematograph, and the ideal woman', *Nineteenth-Century French Studies*, 17, 1 and 2, 1998-99, pp. 71-95; Dave Laing, 'A Voice Without a Face: Popular music and the phonograph in the 1890s', *Popular Music*, 10, 1, 1991, pp. 1-9; Jeffrey Sconce, *Haunted Media: Electronic presence from telegraphy to television*; Anthony Ennis, 'Voices of the Dead: Transmission translation transgression', *Culture, Theory & Critique*, 46, 1, 2005, pp. 11-27.

10. Peter Gibbons, 'The Far Side of the Search for Identity: Reconsidering New Zealand history', *New Zealand Journal of History*, 37, 1, 2003, p. 47.

11. Ibid.

12. Charles Burnett, Michael Fend and Penelope Gouk, eds, *The Second Sense: Studies in hearing and musical judgment from antiquity to the seventeenth century*; James H. Johnson, *Listening in Paris: A cultural history*; Alain Corbin, *Village Bells: Sound and meaning in the 19th century French countryside*; Douglas Kahn, *Noise Water Meat: A history of sound in the arts*; Bruce R. Smith, *The Acoustic World of Early Modern England: Attending to the O-factor*; Shane White and Graham White, 'At Intervals I was Nearly Stunned by the Noise He Made: Listening to African American religious sound in the era of slavery', *American Nineteenth Century History*, 1, 1, 2000, pp. 34-61; Leigh Eric Schmidt, *Hearing Things: Religion, illusion, and the American enlightenment*; Mark M. Smith, *Listening to Nineteenth Century America*; Emily Thompson, *The Soundscape of Modernity: Architectural acoustics and the culture of listening in America*; Richard Cullen Ruth, *How Early America Sounded*; John M. Picker, *Victorian Soundscapes*; Jonathan Sterne, *The Audible Past: Cultural origins of sound reproduction*; Matthew Riley, *Musical Listening in the German Enlightenment: Attention, wonder and astonishment*; Emily Cockayne, *Hubbub: Filth, noise & stench in England 1600-1770*, pp. 106-30; Joy Damousi and Desley Deacon, eds, *Talking and Listening in the Age of Modernity: Essays on the history of sound*; David Toop, *Sinister Resonance: The mediumship of the listener*; Veit Erlmann, *Reason and Resonance: A history of modern aurality*; Kara Keeling and Josh Kun, 'Introduction: Listening to American studies', *American Quarterly Special Issue: Listening to American Studies*, 63, 3, 2011, pp. 445-59; Rey Chow and James A. Steintrager, 'In Pursuit of the Object of Sound', *Differences: A journal of feminist studies, the sense of sound*, 22, 2-3, 2011, pp. 1-9; D. Tracers Scott, 'Sound Studies for Historians of New Media', in David W. Park, Nicholas W. Jankowski and Steve Jones, eds, *The Long History of New Media: Technology, historiography, and contextualising newness*, pp. 75-88. A useful anthology is Mark M. Smith, ed., *Hearing History: A reader*.

13. Keith Sinclair, *A History of New Zealand*; Geoffrey W. Rice, ed., *The Oxford History of New Zealand*; James Belich, *Paradise Reforged: A history of the New Zealanders from the 1880s to the year 2000*; Michael King, *The Penguin History of New Zealand*; Philippa Mein Smith, *A Concise History of New Zealand*; Giselle Byrnes, ed., *The New Oxford History of New Zealand*.
14. Giselle Byrnes, 'Introduction: Reframing New Zealand history', in Byrnes, ed., *New Oxford History of New Zealand*, pp. 2–3.
15. Peter Gibbons, 'Cultural Colonization and National Identity', *New Zealand Journal of History*, 36, 1, 2002, pp. 5–7.
16. Caroline Daley, 'Modernity, Consumption and Leisure', in Byrnes, ed., *New Oxford History of New Zealand*, pp. 423–45.

Overture

1. Jacques Attali, *Noise: The political economy of music*, p. 1.
2. On the physiology of hearing, see R.D. Luce, *Sound and Hearing: A conceptual introduction*. A general introduction to music and neurophysiology is Daniel J. Levitin, *This is Your Brain on Music: The science of a human obsession*. For philosophical perspectives on listening, see Don Ihde, *Listening and Voice: Phenomenologies of sound*; Jean-Luc Nancy, *Listening*; Casey O'Callaghan, *Sounds: A philosophical theory*.
3. Nick Lucas, 'Tip-Toe Through the Tulips', www.youtube.com/watch?v=UZMHJX4b9bU
4. Jason Stanyek and Benjamin Piekut, 'Deadness: Technologies of the intermundane', *TDR: The Drama Review*, 54, 1, 2010, pp. 14–38.
5. *Evening Post*, 14 July 1913, p. 2; *New Zealand Truth*, 24 November 1927, p. 2; Bryan Staff and Sheran Ashley, *For the Record: A history of the recording industry in New Zealand*, p. 29; Dave Cooper, *The Perfect Portable Phonograph*, pp. 14–20. The Decca slogan was based on the English nursery rhyme 'Banbury Cross': 'Ride a cock horse to Banbury Cross/To see a fine lady upon a white horse/With rings on her fingers and bells on her toes/She shall have music wherever she goes'. James Joyce beautifully transmuted this last line into 'Seashell ebb music wayriver she flows' in the poem 'Buy a Book in Brown Paper'. John Gross, ed., *The Oxford Book of Comic Verse*, p. 257.
6. For convenience I have included phonographs under the general term 'gramophones'.
7. For an amusing description of this process, see James H. Johnson, *Listening in Paris: A cultural history*, p. 284.
8. Geoffrey B. Churchman, ed., *Celluloid Dreams: A century of film in New Zealand*, pp. 9–12.
9. Rick Altman, 'The Silence of the Silents', *The Musical Quarterly*, 80, 4, 1996, pp. 648–718; James Lastra, *Sound Technology and the American Cinema: Perception, representation, modernity*, pp. 92–122; Richard Abel and Rick Altman, eds, *The Sounds of Early Cinema*; Rick Altman, *Silent Film Sound*, p. 11.
10. Joan Bulman, *Jenny Lind: A biography*, London, 1956; Steven J. Wurtzler, *Electric Sounds: Technological change and the rise of corporate mass media*, New York,

2007, p. 128; Mark C. Samples, 'The Humbug and the Nightingale: P.T. Barnum, Jenny Lind, and the branding of a star singer for American reception', *Musical Quarterly*, 99, 3/4, 2016, pp. 286–320.

11. Mark Katz, for example, has analysed the changes recording technologies have made to the use of violin vibrato by comparing recorded performances from the early 1900s to the 1940s. He argues that the establishment of vibrato as a normal part of violin playing rather than an ornament was due to the limitations of early recording techniques, coupled with a desire on the part of musicians to add character to their performances. Other musicians imitated the recordings and so vibrato became a standard technique. Mark Katz, *Capturing Sound: How technology has changed music*, pp. 85–89. See also Timothy Day, *A Century of Recorded Music: Listening to musical history*, pp. 142–98; Robert Philip, *Performing Music in the Age of Recording*; Daniel Leech-Wilkinson, 'Recordings and Histories of Performance Style', in Nicholas Cook et al., *The Cambridge Companion to Recorded Music*, pp. 246–62.
12. Katz, Capturing Sound, pp. 72–84.
13. Erika Brady, *A Spiral Way: How the phonograph changed ethnography*, pp. 62–88; Veit Erlmann, 'But What of the Ethnographic Ear?: Anthropology, sound and the senses', in Veit Erlmann, ed., *Hearing Cultures: Essays on sound, listening and modernity*, pp. 1–20; Daniel Makagon and Mark Neumann, *Recording Culture: Audio documentary and the ethnographic experience*, pp. 1–24. For recorded sound and colonialism, see Michael Taussig, *Mimesis and Alterity: A particular history of the senses*, pp. 193–235.
14. Graham Freeman, 'That Chief Undercurrent of My Mind: Percy Grainger and the aesthetics of English folk song', *Folk Music Journal*, 9, 4, 2009, pp. 581–617; Emily Keightley and Michael Pickering, 'For the Record: Popular music and photography as technologies of memory', *European Journal of Cultural Studies*, 9, 4, 2006, pp. 149–65.
15. Deborah Montgomerie, *Love in Time of War: Letter writing in the Second World War*, pp. 59, 68–69.
16. Theodor W. Adorno, 'Opera and the Long Playing Record', in Richard Leppert, ed., *Theodor W. Adorno: Essays on music*, p. 285.
17. Timothy Day, *A Century of Recorded Music: Listening to musical history*, p. 63.
18. Walter Benjamin, 'The Work of Art in the Age of Mechanical Reproduction', in Hannah Arendt, ed., *Illuminations*, pp. 217–51.
19. Theodor .W. Adorno, 'The Form of the Phonograph Record', in Leppert, ed., *Theodor W. Adorno*, p. 278. Adorno also decried the similar effects of radio broadcasts. Adorno, 'Analytical Study of the NBC Music Appreciation Hour', *The Musical Quarterly*, 28, 2, 1994, pp. 325–77; Adorno, 'The Radio Symphony', in Leppert, ed., *Theodor W. Adorno*, pp. 251–70.
20. John Philip Sousa, 'The Menace of Mechanical Music', *Appleton's Magazine*, 8, 1906, pp. 278–84; Constant Lambert, *Music Ho! A Study of Music in Decline*, pp. 168–99.

Chapter 1 Exhibiting Sounds

1. *Evening Post* (*EP*), 19 April 1879, p. 2.
2. *EP*, 19 May 1879, p. 2.
3. *EP*, 16 May 1879, p. 9.
4. John Dix, *Stranded in Paradise: New Zealand rock'n'roll 1955–1988*, pp. 337–41; Bryan Staff and Sheran Ashley, *For the Record: A history of the recording industry in New Zealand*, pp. 11, 150; David Eggleton, *Ready to Fly: The story of New Zealand rock music*, pp. 190–91; Harry Johnson, ed., *Many Voices: Music and national identity in Aotearoa/New Zealand*, pp. 1–9; Chris Bourke, *Blue Smoke: The lost dawn of New Zealand popular music 1919–1964*, pp. 346–50; Glenda Keam and Tony Mitchell, eds, *Home, Land and Sea: Situating music in Aotearoa New Zealand*.
5. Peter Gibbons, 'The Far Side of the Search for Identity', *New Zealand Journal of History*, 37, 1, 2003, p. 47.
6. Claude S. Fischer, 'Gender and the Residential Telephone 1890–1940: Technologies of sociability', *Sociological Forum*, 3, 2, 1988, pp. 211–33; Claude Fischer, *America Calling: A social history of the telephone to 1940*, pp. 222–54; A.C. Wilson, *Wire & Wireless: A history of telecommunications in New Zealand 1890–1987*, pp. 78–107; David Mercer, *The Telephone: The life story of a technology*, pp. 39–56.
7. Lisa Gitelman has noted that analyses of the production of sound cannot be reduced to the activities of individuals such as Edison or corporations, and that 'the phonograph provides an exemplary instance of cultural production snatched from the hands of putative producing agents'. See Lisa Gitelman, 'How Users Define New Media: A history of the amusement phonograph', in David Thorburn and Henry Jenkins, eds, *Rethinking Media Change: The aesthetics of transition*, p. 74.
8. Oliver Read and Walter L. Welch, *From Tin Foil to Stereo: Evolution of the phonograph*, pp. 11–57; Roland Gelatt, *The Fabulous Phonograph 1877–1977*, pp. 1–32; V.K. Chew, *Talking Machines*, pp. 5–8; Ross Laird, *Sound Beginnings: The early record industry in Australia*, pp. 1–3; Marsha Seifert, 'Aesthetics, Technology and the Capitalization of Culture: How the talking machine became a musical instrument', *Science In Context*, 8, 2, 1995, pp. 417–40, esp. pp. 422–25; René Rondeau, *Tinfoil Phonographs: The dawn of recorded sound*, pp. 11–14; Lisa Gitelman, 'Souvenir Foils: On the status of print at the origin of recorded sound', in Lisa Gitelman and Geoffrey Rice, eds, *New Media 1740–1915*, pp. 157–73.
9. *Daily Southern Cross*, 18 June 1866, p. 1; 13 April 1871, p. 1; *North Otago Times* (*NOT*), 15 November 1872, p. 6; *Waikato Times*, 20 September 1873, p. 3; *Otago Witness* (*OW*), 19 February 1876, p. 7.
10. *Nelson Examiner and New Zealand Chronicle*, 20 November 1861, p. 4. For the phonautograph, see Read and Welch, *From Tin Foil*, pp. 3–6; Rondeau, *Tinfoil Phonographs*, p. 18; Jonathon Sterne, *The Audible Past: Cultural origins of sound reproduction*, pp. 45–51. For an account of recovering audio from a phonautograph and a recording, see 'World's Oldest Recording Made Available Online', *First Sounds*, 27 March 2008, www.firstsounds.org

11. *Wanganui Herald* (*WH*), 23 February 1878, p. 2; *NOT*, 6 April 1878, p. 2; *EP*, 6 April 1878, p. 1; *OW*, 18 May 1878, p. 19; *Taranaki Herald*, 12 June 1878, p. 2.
12. *Marlborough Express*, 10 April 1879, p. 4.
13. Gitelman, 'Souvenir Foils', p. 157.
14. *Wanganui Herald*, 20 December 1879, p. 2.
15. Wally Golledge, 'Early Days of the Phonograph in Nelson', *The Phonographic Record: Supplement*, 4, 5, 1969, p. 2.
16. Walter Norris, 'The Edison Phonograph in Christchurch Between 1879 and 1914', *The Phonographic Record*, 8, 4, 1973, p. 80.
17. *EP*, 19 April 1879, p. 2.
18. *EP*, 17 May 1879, p. 2.
19. *EP*, 19 May 1879, p. 5.
20. *EP*, 17 May 1879, p. 6.
21. *OW*, 31 July 1880, p. 20; *NOT*, 19 October 1881, p. 2; *EP*, 27 December 1881, p. 3.
22. *EP*, 22 August 1887, p. 2; *OW*, 21 October 1887, p. 10.
23. *OW*, 26 September 1889, p. 35.
24. Gelatt, *The Fabulous Phonograph*, pp. 36–57.
25. Laird, *Sound Beginnings*, pp. 3–9; Henry Reese, '"The World Wanderings of a Voice": Exhibiting the cylinder phonograph in Australasia', in Joy Damousi and Paula Hamilton, eds, *A Cultural History of Sound, Memory, and the Senses*, pp. 25–39.
26. *OW*, 8 January 1891, p. 8; *Press*, 13 January 1891, p. 4; *Timaru Herald*, 7 March 1891, p. 2; *West Coast Times*, 7 May 1891, p. 4.
27. *Otago Daily Times* (*ODT*), 17 December 1890, p. 3.
28. *ODT*, 31 December 1890, p. 2.
29. *NZH*, 16 February 1891, p. 5; *Auckland Weekly News*, 21 February 1891, p. 37; *New Zealand Graphic*, 28 February 1891, p. 8; 14 March 1891, pp. 7–8; Donald Kerr, *Amassing Treasures for All Times: Sir George Grey, colonial bookman and collector*, p. 253. The recordings by Grey have been lost.
30. *OW*, 8 January 1891, p. 28; *Auckland Weekly News*, 7 March 1891, p. 23; *West Coast Times*, 7 May 1891, p. 4.
31. *New Zealand Graphic*, 17 January 1891, p. 12.
32. *Sporting Review*, 7 February 1891, p. 5.
33. *Sporting Review*, 14 February 1891, p. 7.
34. *Press*, 11 January 1891, p. 6.
35. Gelatt, *The Fabulous Phonograph*, p. 29.
36. *EP*, 13 May 1891, p. 2; *OW*, 28 May 1891, p. 3; Herman Klein, 'Sims Reeves: Prince of English tenors', in Roger Wimbush, ed., *The Gramophone Jubilee Book 1923–1973*, pp. 109–12.
37. *Press*, 13 January, 1891, p. 5; 15 January 1891, p. 3.
38. *New Zealand Graphic and Ladies' Journal* (*NZG*), 7 March 1891, p. 9.
39. Ibid.
40. *Te Aroha News*, 19 October 1889, p. 3; *Waimate Daily Advertiser*, 8 November 1900, p. 1; *New Zealand Free Lance*, 27 December 1902, p. 26; *OW*, 13 August 1902, p. 78.

41. *ODT*, 17 December 1890, p. 3.
42. *New Zealand Observer*, 14 February 1891, p. 7.
43. *EP*, 18 March 1891, p. 2.
44. Patricia Jalland, *Death in the Victorian Family*, pp. 1–16; Judith Flanders, *The Victorian House: Domestic life from childbirth to deathbed*, pp. 302–48.
45. *The New Phonogram*, February 1912, p. 18, April 1912, pp. 12–13, August 1912, p. 18; The National Phonograph Company, *The Phonograph and How To Use It*, pp. 173–78, 143–44.
46. James Joyce, *Ulysses*, p. 109.
47. *ODT*, 17 December 1890, p. 3; *Press*, 13 January 1891, p. 4; *New Zealand Observer*, 14 February 1891, p. 7; *EP*, 14 March 1891, p. 2.
48. *North Otago Times*, 22 January 1892, p. 3; 21 March 1893, p. 3; *OW*, 13 April 1893, p. 37; *Hawera & Normanby Star*, 29 March 1894, p. 2; 31 July 1894, p. 2; *EP*, 9 February 1898, p. 2.
49. *EP*, 9 February 1898, p. 4; 10 February 1898, p. 6; 9 February 1898, p. 4.
50. *Grey River Argus*, 18 July 1898, p. 2; *Wanganui Herald*, 1 September 1898, p. 2.
51. *EP*, 25 September 1899, p. 5.

Chapter 2 Selling Sounds

1. Ruth Edge, 'Mr Bohanna Goes Down Under: Parts 1–3', *Australian Record and Music Review*, 7, 1990, pp. 12–15; 10, 1991, pp. 12–16; 17, 1993, pp. 3–5; Ross Laird, *Sound Beginnings: The early record industry in Australia*, pp. 89–92; Bryan Staff and Sheran Ashley, *For the Record: A history of the recording industry in New Zealand*, p. 21.
2. Geoffrey Jones, 'The Gramophone Company: An Anglo-American multinational 1898–1931', *Business History Review*, 59, 1, 1985, p. 81.
3. Ali Jihad Racy, 'Record Industry and Egyptian Traditional Music: 1904–1932', *Ethnomusicology*, 20, 1, 1976, pp. 23–48; Pekka Gronow, 'The Record Industry Comes to the Orient', *Ethnomusicology*, 25, 2, 1981, pp. 251–84; Gerry Farrell, 'The Early Days of the Gramophone Industry in India: Historical, social and musical perspectives', *British Journal of Ethnomusicology*, 3, 1993, pp. 31–52; Reebee Garofalo, 'From Music Publishing to MP3: Music and industry in the twentieth century', *American Music*, 17, 3, 1999, pp. 323–29; Andrew F. Jones, *Yellow Music: Media culture and colonial modernity in the Chinese jazz age*, pp. 53–72.
4. Staff and Ashley, *For the Record*, p. 21.
5. Marsha Siefert, 'The Audience at Home: The early recording industry and the marketing of musical taste', in James S. Ettema and D. Charles Whitney, eds, *Audiencemaking: How the media create the audience*, pp. 186–214; Emily Thompson, 'Machines, Music, and the Quest for Fidelity: Marketing the Edison phonograph in America 1877–1925', *The Musical Quarterly*, 79, 1, 1995, pp. 131–71.
6. Roland Gelatt, *The Fabulous Phonograph 1877–1977*, pp. 114–29; Marsha Siefert, 'Aesthetics, Technology, and the Capitalization of Culture: How the talking machine became a musical instrument', *Science In Context*, 8, 2, 1995,

pp. 417–49; Thompson, 'Machines, Music and the Quest for Fidelity', pp. 131–71; William Kenney, *Recorded Music in American Life: The phonograph and popular memory 1890–1945*, pp. 52–64; Timothy C. Fabrizio and George E. Paul, *Antique Phonograph Advertising: An illustrated history*, pp. 65, 91, 96, 122–23, 158; Mark Katz, *Capturing Sound: How technology has changed music*, pp. 48–61.
7. For details of these and other performers, see Tim Gracyk and Frank Hoffmann, *Popular American Recording Pioneers: 1895–1925*. Many recordings of these performers can be heard and downloaded at the University of California Santa Barbara cylinder preservation collection website. UCSB Cylinder Audio Archive, http://cylinders.library.ucsb.edu/
8. G.P. Walsh, 'Narelle, Marie (Molly) (1870–1941)', *Australian Dictionary of Biography*, National Centre of Biography, Australian National University, http://adb.anu.edu.au/biography/narelle-marie-molly-13126/text23753
9. *Wairarapa Age*, 9 January 1907, p. 6.
10. *Evening Post (EP)*, 24 November 1905, p. 8.
11. René Rondeau, 'The Victor Auxetophone', *In the Groove*, 25, 9, 2000, pp. 4–6, 14–15.
12. *Dominion*, 2 February 1908, p. 1.
13. *Dominion*, 3 February 1908, p. 4. Both 'gramaphone' and 'gramophone' were used during the early twentieth century.
14. *EP*, 23 December 1908, p. 8.
15. Charles Begg & Sons, *Begg's Jubilee Souvenir 1861–1911*, pp. 20, 24, 26, 28.
16. Ibid., p. 20.
17. John Philip Sousa, 'The Menace of Mechanical Music', *Appleton's Magazine*, 8, 1906, p. 282.
18. Ibid., pp. 278–84, cited in Patrick Feaster, 'Phonozoic Text Archive, Document 155', *Phonozoic*, n.d., www.phonozoic.net/n0155.htm
19. *Wanganui Herald*, 2 November 1906, p. 4.
20. *Otago Witness (OW)*, 12 September 1906, p. 58.
21. *EP*, 24 September 1909, p. 8.
22. Andre Millard, *America on Record: A history of recorded sound*, p. 59.
23. Rollo Arnold, 'The Country Child in Later Victorian New Zealand', in H. Debenham and W. Slinn, eds, *Australasian Victorian Studies Association: Conference papers*, pp. 1–15; John MacGibbon, *Piano in the Parlour: When the piano was New Zealand's home entertainment centre*, pp. 28–59.
24. James Parakilas, ed., *Piano Roles: A new history of the piano*, pp. 150–261.
25. 'Lorenski', 'Shut That Old Piano Up: Written for the new phonograph', *The New Phonogram*, 2, 2, 1910, pp. 10–11.
26. Graham Melville-Mason, 'The Gramophone as Furniture', in Alistair G. Thomson, ed., *Phonographs & Gramophones: A symposium organised by the Royal Scottish Museum in connection with the exhibition Phonographs and Gramophones and the Centenary of the Invention of the Phonograph by Thomas Alva Edison*, pp. 117–38; Holly Kruse, 'Early Audio Technology and Domestic Space', *Stanford Humanities Review*, 3, 2, 1993, pp. 1–14; Timothy C. Fabrizio, 'Ordinary People: The talking machine in real life', *ARSC Journal*, 30, 1, 1999,

pp. 20–25; Kyle S. Barnett, 'Furniture Music: The phonograph as furniture 1900–1930', *Journal of Popular Music Studies*, 18, 3, 2006, pp. 301–24.
27. Melville-Mason, 'The Gramophone as Furniture', p. 119.
28. Compton MacKenzie, 'The Gramophone: Its past, its present, its future', *Proceedings of the Musical Association*, 51st Sess., London, 1924–1925, p. 100.
29. H.L. Wilson, *Music and the Gramophone and Some Master Recordings*, p. 7.
30. Barnett, 'Furniture Music', p. 303.
31. *Auckland Weekly News*, 11 September 1913, p. 86.
32. *New Zealand Observer*, 16 March 1901, p. 20; 11 July 1915, p. 21; *Auckland Weekly News*, 29 January 1903, p. 51; 4 April 1907, p. 58; *EP* 10 August 1911, p. 6.
33. *Appendices to the Journals of the House of Representatives*, 1901, E–1, p. xii; 1920, E–1, p. 33.
34. MacGibbon, *Piano in the Parlour*, p. 68; Kirstine Moffat, *Piano Forte: Stories and sounds from colonial New Zealand*, pp. 207–08.
35. MacGibbon, *Piano in the Parlour*, pp. 70–71.

Chapter 3 Domesticating Sounds

1. Kirstine Moffat, *Piano Forte: Stories and soundscapes from colonial New Zealand*, pp. 68–74.
2. William H. Kenney, *Recorded Music in American Life: The phonograph and popular memory 1890–1945*, pp. 88–108.
3. 'Edison Grand Opera Amberol Records', *The New Phonogram*, 2, 8, 1910, p. 12.
4. Marc Perlman, 'Golden Ears and Meter Readers: The contest for epistemic authority in audiophilia', *Social Studies of Science*, 34, 5, 2004, pp. 783–807.
5. Timothy C. Fabrizio and George F. Paul, *Antique Phonograph Advertising: An illustrated history*, pp. 125, 135, 136; Mark Katz, *Capturing Sound: How technology has changed music*, p. 58; *New Zealand Observer* (*NZO*), 16 March 1901, p. 20; *Auckland Weekly News* (*AWN*), 16 November 1905, p. 5; *Evening Post* (*EP*), 25 September 1909, p. 8; *AWN*, 11 September 1913, p. 86; *Otago Daily Times* (*ODT*), 2 June 1924, p. 3.
6. *NZO*, 16 March 1901, p. 20. See also *AWN*, 20 June 1907, p. 33.
7. *Wairarapa Age*, 23 January 1907, p. 3.
8. T.A. Moresby, 'Music in Paeroa from 1908 to 1928', *Ohinemuri Regional History Journal*, 6, 1, 1969, p. 20.
9. Vincent O'Sullivan and Margaret Scott, eds, *The Collected Letters of Katherine Mansfield*, p. 28.
10. 'The Edison Phonograph as a Home Educator', *The New Phonogram*, 3, 2, 1911, p. 15.
11. Examples of the many such guides include: Anne Shaw Faulkner, *What We Hear in Music* (published by the Victor Talking Machine Company); *Opera at Home* (published by the Gramophone Company); Percy A. Scholes, *Learning to Listen by Means of the Gramophone* (published by the Gramophone Company).
12. Abigail Cooke, 'Humorous Reflections on Laughing Records', *ARSC Journal*, 32, 2, 2001, pp. 232–42; Jacob Smith, 'The Frenzy of the Audible: Pleasure,

authenticity and recorded laughter', *Television & New Media*, 6, 1, 2005, pp. 23–47.
13. P.G. Wodehouse, *Psmith in the City* (1910), London, 1971, p. 21.
14. *New Zealand Free Lance* (*NZFL*), 22 September 1906, p. 2; 18 January 1908, p. 17.
15. *Otago Witness*, 18 January 1879, p. 24.
16. *Te Aroha News*, 28 July 1888, p. 3.
17. *NZFL*, 10 May 1902, p. 22.
18. *NZFL*, 27 December 1902, p. 6.
19. 'In a Phonograph Store', *The New Phonogram*, 4, 2, 1912, pp. 10–11.
20. C.E. Woledge, 'Talking Machine Memories', *The Phonographic Record*, 2, 2, 1966, pp. 15–16.
21. *Poverty Bay Herald*, 23 June 1915, p. 6; *Ashburton Guardian*, 25 August 1915, p. 6; *EP*, 25 February 1916, p. 3; 15 August 1916, p. 5.
22. *Chronicles of the NZEF*, 1, 1, August 1916, p. 12; 2, 15, 1917, p. 58; 4, 43, 1918, p. 156.
23. *Chronicles of the NZEF*, 1, 1, 30 August 1916, p. 12.
24. Archibald Baxter, *We Will Not Cease*, p. 57.
25. For the roles of print culture in the global recording industry, see Colin Symes, *Setting the Record Straight: A material history of classical recording*.

Chapter 4 Teaching Sounds

1. Olive Boyd, interview by Jamie Mackay, 7 February 1992, for the Huntly Coalfields Oral History Project, Oral History Centre, Alexander Turnbull Library, OHInt-0020/03.
2. This technique produces a faint but clear sound that emphasises the voice.
3. Thomas Edison, 'The Phonograph and its Future', *North America Review*, 136, 1878, pp. 527–32; 'The Perfected Phonograph', *North America Review*, 166, 1888, pp. 641–50.
4. See, for example, Percy Scholes, *The Listener's History of Music: A book for any concert-goer, gramophonist or radio listener, providing also a course of study for adult classes in the appreciation of music*; *Practical Lesson Plans in Musical Appreciation by Means of the Gramophone*; *Music, the Child and the Masterpiece: A comprehensive handbook of aims and methods in all that is usually called 'musical appreciation'*; G.N. Sharp, 'The Gramophone in Musical Education', *Music & Letters*, 19, 2, 1938, pp. 199–202. An account of one of the main 'boosters' of the educational gramophone is H.C. Colles, *Walford Davies: A biography*.
5. *Maoriland Worker*, 29 August 1923, p. 9; *Evening Post*, 6 October 1928, p. 24; *Auckland Star*, 12 December 1929, p. 12.
6. See Appendix 1 in William W. Johnson, *The Gramophone in Education: An introduction to its use in school and in the home*, pp. 145–49.
7. Ibid., p. 42.
8. Colin Symes, 'A Sound Education: The gramophone and the classroom in the United Kingdom and the United States 1920–1940', *British Journal of Music*

Education, 21, 2, 2004, pp. 166–67. See also Mark Katz, *Capturing Sound: How technology has changed music*, pp. 48–71.
9. Susan Braatvedt, 'A History of Music Education in New Zealand State Primary and Intermediate Schools 1878–1989', pp. 144–91; Braatvedt, 'Douglas Tayler: Man of action and initiative', *Sound Ideas*, 6, 1, 2003, pp. 6–18.
10. Douglas Tayler, *A Complete Scheme of School Music Related to Human Life*, p. 18.
11. Douglas Tayler, 'The Education Department and School Music', *Music in New Zealand*, 1, 1, 1931, p. 14.
12. Tayler, *A Complete Scheme*, p. 18. See also pp. 25–26, 34–35, 43, 50 for notes on appropriate use of the gramophone for each level from Junior to Senior.
13. Ibid., pp. 67–100.
14. Ibid., p. 71.
15. Ibid., p. 78.
16. Ibid., p. 97.
17. Ibid., p. 25.
18. Department of Education, *Syllabus of Instruction for Public Schools*, p. 195.
19. Susan Braatvedt, 'A Few Musical Memories', *Sound Ideas*, 6, 1, 2003, p. 55.
20. Margaret Archer, 'Margaret Coyne (nee Archer)', in Ken Clover, ed., *The People of the Plains: Some early memories of those who have lived in or been associated with the Hauraki Plains Volume 1*, p. 246.
21. Caroline Daley, *Leisure & Pleasure: Reshaping and revealing the New Zealand body 1900–1960*, pp. 226–31.
22. 'Myrtle Fraser-Jones', in Clover, ed., *The People of the Plains Volume 2*, p. 325.
23. A.N. Fitzgerald, 'New Zealand Speech: Impressions and reflections', *Education Gazette*, 13, 1934, p. 123.
24. Janet Soler, 'Renegotiating Cultural Authority: Imperial culture and the New Zealand primary school curriculum in the 1930s', *History of Education*, 35, 1, 2006, p. 18.
25. Tayler, *A Complete Scheme*, p. 36.
26. Johnson, *The Gramophone in Education*, p. 62.
27. Philip Norman, *Douglas Lilburn: His life and music*, p. 37.
28. Rachel Barrowman, *Victoria University of Wellington 1899–1999: A history*, p. 59.
29. Norman, *Douglas Lilburn*, p. 61. See also John M. Jennings, *Music at Canterbury: A centennial history of the School of Music, University of Canterbury, Christchurch, New Zealand, 1891–1991*, pp. 25, 26.
30. Johnson, *The Gramophone in Education*, p. 120.
31. Norman, *Douglas Lilburn*, pp. 58, 73.
32. Ibid., p. 58.
33. Ibid., p. 63.
34. Roy Shuker, *Educating the Workers?: A history of the Workers' Education Association in New Zealand*, pp. 68–69.
35. Ibid., p. 93.
36. Ibid., pp. 93–94.
37. See Rachel Hawkey, 'This is a British Colony: Vernon Griffiths and music education in New Zealand', *Music in New Zealand*, 14, 1991, pp. 30–36; Hawkey, 'Vernon Griffiths: His life and philosophy of music education as demonstrated

through his collected papers'; Hawkey, 'A Man of Missionary Zeal: Vernon Griffiths', *Sound Ideas*, 6, 1, 2003, pp. 19–25; Hawkey, 'Griffiths, Thomas Vernon 1894–1985', Dictionary of New Zealand Biography, www.teara.govt.nz/en/biographies/4g21/1

38. Hawkey, 'Griffiths, Thomas Vernon 1894–1985'.
39. For film societies, see Simon Sigley, 'Film Culture: Its development in New Zealand 1929–1972'; Sigley, 'How *The Road to Life* (1931) Became the Road to Ruin: The case of the Wellington Film Society in 1933', *New Zealand Journal of History*, 42, 2, 2008, p. 198. For the activities of literary elite groups, see Chris Hilliard, *The Bookmen's Dominion: Cultural life in New Zealand 1920–1950*; Rachel Barrowman, *A Popular Vision: The arts and the left in New Zealand 1930–1950*, pp. 61–88.
40. *Music in New Zealand*, 3, 6, 1933, p. 7.
41. *Music in New Zealand*, 3, 8, 1933, p. 7.
42. *Music in New Zealand*, 4, 10, 1935, p. 13.
43. *Music in New Zealand*, 4, 12, 1934, p. 11.
44. *Music in New Zealand*, 5, 5, 1935, p. 14.
45. *Pianoforte and Gramophone Recital*, c. 1923; *Dominion*, 24 April 1926, p. 20; 18 June 1927, p. 20; *Moore's Gramophone Programme*, 5 June 1929.
46. *Dominion*, 24 April 1926, p. 20.
47. Gavin East, 'Edison Diamond Discs: New Zealand perspective', *The Phonographic Record*, 36, 2, 2001, p. 39.
48. The Turing Test, originally developed by the British mathematician Alan Turing, is often used as a criterion for assessing the possibility of Artificial Intelligence. Person A communicates via text with person B and a software program. If person A cannot tell which is which, then the program might be described as 'intelligent'.
49. For accounts of tone tests, see Oliver Read and Walter L. Welch, *From Tin Foil to Stereo: Evolution of the phonograph*, pp. 189–204; Emily Thompson, 'Machines, Music and the Quest for Fidelity: Marketing the Edison phonograph in America 1877–1925', *The Musical Quarterly*, 79, 1, 1995, pp. 131–71; Jonathan Sterne, *The Audible Past: Cultural origins of sound reproduction*, pp. 261–65; Steve J. Wurtzler, *Electric Sounds: Technological change and the rise of corporate mass media*, pp. 80–81.
50. Laurie Lewis, *Arthur and the Nights at the Turntable: The life and times of a jazz broadcaster*; Chris Bourke, *Blue Smoke: The lost dawn of New Zealand popular music 1918–1964*, pp. 70–71.
51. Lewis, *Arthur and the Nights at the Turntable*, p. 46.
52. Gordon Spittle, *Counting the Beat: A history of New Zealand song*, p. 11.
53. Alan Turley, 'The Incredible Tex Morton', *New Zealand Memories*, 37, 2002, pp. 4–12; Gordon Spittle, 'Morton, Tex 1916–1983', Dictionary of New Zealand Biography, www.teara.govt.nz/en/biographies/5m59/1; Spittle, *The Tex Morton Songbook*, pp. 3–115; Bourke, *Blue Smoke*, pp. 81–85.
54. Gordon Spittle, 'Johnny Cooper: The original Maori cowboy', *Music in New Zealand*, 24, 1994, p. 43; Bourke, *Blue Smoke*, pp. 246–47, 279–80.

55. Colin Dorsey, *Dance Bands of the 20th Century in North Otago*, p. 64.
56. Katz, *Capturing Sound*, p. 78. See pp. 73–84 for a general discussion of jazz learning and performance as affected by recorded sound.

Chapter 5 Moving Sounds

1. Roland Gelatt, *The Fabulous Phonograph 1877-1977*, pp. 265–77; Jerrold Northrop Moore, *Sound Revolutions: A biography of Fred Gaisberg, founding father of commercial sound recording*, pp. 287–96.
2. *Music in New Zealand*, 3, 7, 1933, p. 11; 3, 8, 1933, p. 7.
3. Buick reviewed classical concerts and wrote a book about Felix Mendelssohn's oratorio *Elijah*. See J.E. Traue, 'Buick, Thomas Lindsay 1865–1938', Dictionary of New Zealand Biography, www.teara.govt.nz/en/biographies/3b57/1
4. C.S. Davis, 'Service Without Sales: The dealer's point of view again', *Gramophone*, 6, 72, 1929, pp. 526–28, cited in D.L. LeMahieu, 'The Gramophone: Recorded music and the cultivated mind in Britain between the wars', *Technology and Culture*, 23, 3, 1982, p. 382.
5. Brian Salkeld, 'The Dancing Decade: 1920–1930', *Stout Centre Review*, 2, 4, 1992, pp. 3–12; John Griffiths, 'Popular Culture and Modernity: Dancing in New Zealand society 1920–1945', *Journal of Social History*, 41, 13, 2008, pp. 611–32; Chris Bourke, *Blue Smoke: The lost dawn of New Zealand popular music 1918-1964*, pp. 39–43, 50–93.
6. Bourke, *Blue Smoke*, pp. 40–43.
7. *New Zealand Truth*, 14 March, 1929, p. 8; 21 March 1929, p. 8.
8. *New Zealand Herald*, 25 March 1929, p. 7.
9. Mary Findlay, *Tooth and Nail: The story of a daughter of the Depression*, p. 154.
10. Laurie Lewis, *Arthur and the Nights at the Turntable: The life and times of a jazz broadcaster*, pp. 81–82.
11. Deborah Montgomerie, *Love in Time of War: Letter writing in the Second World War*, pp. 58, 68.
12. 'Elsie Billings' in Ken Clover, ed., *The People of the Plains: Some early memories of those who have lived in or been associated with the Hauraki Plains Volume 1*, p. 48.
13. Douglas Tayler, 'A Musical Query', *Education Gazette*, 1 May 1928, p. 67.
14. Karl Atkinson, 'Jazz: A musical upstart drenching the world with din', *The Mirror*, 1 February 1931, p. 18.
15. Karin Bijesterveld, 'The Diabolical Symphony of the Mechanical Age: Technology and symbolism of sound in European and North American noise abatement campaigns 1900–1940', *Social Studies of Science*, 31, 1, 2001, p. 53.
16. Jessie Annett-Wood, 'The Modern Woman in *The Mirror*: Modernity and the New Zealand women's magazine, 1922–1932'.
17. Atkinson, 'Jazz', pp. 17–18.
18. F.M. Price, 'Jazz: A musical retrogression', *Music in New Zealand*, 4, 2, 1934, pp. 3–4.
19. See, for example Scott DeVeaux, 'Constructing the Jazz Tradition: Jazz historiography', *Black American Literature Forum*, 25, 3, 1991, pp. 525–60; Carol

J. Oja, 'Gershwin and American Modernists of the 1920s', *The Musical Quarterly*, 78, 4, 1994, pp. 646–68; John Baxendale, 'Into Another Kind of Life in Which Anything Might Happen: Popular music and late modernity 1910–1930', *Popular Music*, 14, 2, 1995, pp. 137–54; Paul Lopes, 'Diffusion and Syncretism: The modern jazz tradition', *Annals of the American Academy of Political and Social Science*, 566, 1999, pp. 25–36; Bruce Johnson, *The Inaudible Music: Jazz, gender and Australian modernity*; Andrew Jones, *Yellow Music: Media culture and colonial modernity in the Chinese jazz age*; Joel Dinerstein, *Swinging the Machine: Modernity, technology, and African American culture between the two world wars*; Alexander Weheliye, *Phonographies: Grooves in sonic Afro-modernity*; Jonathan Wipplinger, 'The Aural Shock of Modernity: Weimar's experience of jazz', *The Germanic Review*, 82, 4, 2007, pp. 299–320.
20. 'Myrtle Fraser-Jones', in Clover, *The People of the Plains Volume 2*, p. 329.
21. Allison McCracken, 'God's Gift to Us Girls: Crooning, gender and the re-creation of American popular song 1928–1933', *American Music*, 17, 4, 1999, pp. 365–88; Paula Lockhart, 'A History of Early Microphone Singing 1925–1939: American mainstream popular singing at the advent of the electronic microphone amplification', *Popular Music and Society*, 26, 3, 2003, pp. 367–85.
22. Allison McCracken, 'Real Men Don't Sing Ballads, The Radio Crooner in Hollywood', in Pamela Robertson Wojcik and Arthur Knight, eds, *Soundtrack Available: Essays on film and popular music*, pp. 105–34. See also the radio debate between Arthur Pearse and Stanley Oliver, head of Wellington's Harmonic Society, transcribed in Lewis, *Arthur and the Nights at the Turntable*, pp. 53–67; *Press*, 24 April 1937, p. 11.
23. Jock Phillips, *A Man's Country? The Image of the Pakeha Male: A history*, pp. 217–60.
24. John Philip Sousa, 'The Menace of Mechanical Music', *Appleton's Magazine*, 8, 1906, pp. 278–84, cited in Patrick Feaster, 'Phonozoic Text Archive, Document 155', *Phonozoic*, n.d., www.phonozoic.net/n0155.htm
25. Constant Lambert, *Music Ho! A Study of Musical Decline*, pp. 163–93. For Adorno see, in particular, Theodor W. Adorno, 'The Form of the Phonograph Record' (1934), in *Essays on Music*, pp. 277–82; and 'On Popular Music' (1941), in ibid., pp. 437–69.
26. See, for example, Julius Lemmer, 'The Amateur in Music', *Music in New Zealand*, 1, 2, 1931, p. 33; Isabell M. Cluett, 'The Menace of Mechanical Art', *The Mirror*, July 1931, pp. 32–33; Elizabeth Connors, 'Mechanised Pleasure: Why we get bored', *The Mirror*, September 1931, p. 73.
27. Cluett, 'Mechanical Art', p. 32.
28. Connors, 'Mechanised Pleasure', p. 73.
29. Lambert, *Music Ho!*, p. 164.
30. Ibid., p. 238.
31. Buick to Stewart, 5 September 1927, Bruce Anderson Papers, MS-0351, Alexander Turnbull Library, Wellington.
32. Jill Julius Matthews, *Dance Hall & Picture Palace: Sydney's romance with modernity*, pp. 127–42.

33. Gordon Spittle, 'Morton, Tex 1916–1983', Dictionary of New Zealand Biography, www.teara.govt.nz/en/biographies/5m59/1; Bourke, *Blue Smoke*, p. 82.
34. Claudia Lemke, 'Maori Involvement in Sound Recording and Broadcasting 1919 to 1958', pp. 83–88.
35. Pekka Gronow, 'The Record Industry Comes to the Orient', *Ethnomusciology*, 25, 2, 1981, pp. 274–75.
36. The performers were: Ana Hato, Ani Patiti, Polly Tonihi, Marie Mihaka, Tina Hunt, Mary Amohau, Minnie Eparakama, Deane Waretini, William Hunt, Tai Amohau and Molly Wilson, pianist. *Evening Post*, 30 April 1927, p. 4.
37. Alan Armstrong, 'The Great Songs of Ana Hato and Deane Waretini', *Te Ao Hou*, 44, 1963, pp. 57–58; Walter Norris, 'Ana Hato and Deane Waretini', *The Phonographic Record*, 2, 5, 1967, pp. 36–37; Alison Masters, 'Ana Hato 1906–1953', in Charlotte Macdonald, Merimeri Penfold and Bridget Williams, eds, *The Book of New Zealand Women: Ko Kui Ma te Kaupapa*, pp. 277–78; Jonathan Dennis, 'Ana Hato: The melody is ended but the memory lingers on', *Music in New Zealand*, 31, 1995/6, pp. 46–47; Sleevenotes, *Ana Hato*, CD, Wellington, 1996; Joe Malcolm, 'Hato, Ana Matawharua 1907–1953, Dictionary of New Zealand Biography, www.teara.govt.nz/en/biographies/4h19/1; Bourke, *Blue Smoke*, pp. 33–34; Amanda Mills, 'Ana Hato: Profile', www.audioculture.co.nz/people/ana-hato
38. *Evening Post* (*EP*), 11 June 1927, p. 24.
39. Masters, 'Ana Hato', p. 277.
40. Atholl Anderson, Judith Binney and Aroha Harris, *Tangata Whenua: An illustrated history*, p. 376.
41. Parlophone promotional leaflet quoted in CD sleevenotes to *Ana Hato*.
42. For an overview, see Gronow, 'The Record Industry'; Ali Jihad Racy, 'Record Industry and Egyptian Traditional Music: 1904–1922', *Ethnomusicology*, 20, 1, 1976, pp. 23–48; Gerry Farrell, 'The Early Days of the Gramophone Industry in India: Historical, social and musical perspectives', *British Journal of Ethnomusicology*, 2, 1993, pp. 31–53.
43. Anon., 'The Singing Tahiwis', *Otaki Historical Society Journal*, 21, 1998, pp. 45–49; Sleevenotes, *Te Whanau Tahiwi*, CD, Wellington, 1998; Bourke, *Blue Smoke*, p. 35.
44. Sleevenotes to *Te Whanau Tahiwi*.
45. Parlophone brochure quoted in ibid.
46. *EP*, 21 July 1930, p. 23.
47. Michael Alexandratos, 'The Rotorua Maori Choir: A history, discography and analysis, *The Discographer Magazine*, 3, 6, 2016, pp. 3–24; Michael Alexandratos, 'Rotorua Maori Choir: Profile', www.audioculture.co.nz/people/rotorua-maori-choir
48. Peter Downes, 'Dech, Gil 1897–1974', Dictionary of New Zealand Biography, www.teara.govt.nz/en/biographies/5d13/1. See also *Dominion*, 10 May 1930, p. 22; *The Mirror*, 1 June 1930, p. 58; *Dominion*, 28 June 1930, p. 20; Allen Armstrong, 'Still Popular After Thirty Years: How Rotorua club made first recordings of Maori music', *Te Ao Hou*, 36, 1961, pp. 63–64; Bruce Anderson,

The EMI NZ Story, 82-011-02, Alexander Turnbull Library, Wellington; Bourke, *Blue Smoke*, p. 34.
49. Benedict Anderson, *Imagined Communities: Reflections on the origin and spread of nationalism*, pp. 5-7.
50. Lemke, 'Maori Involvement in Sound Recording and Broadcasting', p. 3; Pat Hohepa, 'Maori and Pakeha: The one-people myth', in Michael King, ed., *Tihe Mauri Ora: Aspects of Maoritanga*, pp. 98-111.
51. Tairawhiti Maori Association, *Echoes of the Pa: Research proceedings 1931-2*, p. 9.
52. Christopher Burns, 'Parading Kiwis: New Zealand soldier concert parties, 1916-1954', passim.
53. No sales figures are available. For reviews, see *Dominion*, 1 September 1928, p. 22; 28 September 1928, p. 24; 15 September 1930, p. 25; *New Zealand Herald*, 22 September 1928, p. 8.
54. See Ernest McKinlay, *Ways and By-Ways of a Singing Kiwi*, p. 156.
55. Fred Gaisberg, *Music on Record*, p. 52.

Chapter 6 Receiving Radio

1. *Evening Post* (*EP*), 14 August 1913, p. 10.
2. *New Zealand Statutes* (*NZS*), 1903, p. 28.
3. *EP*, 14 August 1913, p. 10.
4. *Ashburton Guardian*, 14 August 1913, p. 7; *Feilding Star*, 14 August 1913, p. 3; *Hawera & Normanby Star*, 14 August 1913, p. 4; *Wanganui Chronicle*, 14 August 1913, p. 4.
5. Asa Briggs and Peter Burke, *A Social History of Media: From Gutenberg to the internet*, pp. 155-58.
6. Hillel Schwartz, *Making Noise: From Babel to the big bang and beyond*, pp. 408-17.
7. Carolyn Marvin, *When Old Technologies Were New*, p. 8.
8. I will use 'wireless' throughout this chapter to distinguish it from broadcast radio.
9. Howard Robinson, *A History of the Post Office in New Zealand*, pp. 182-87, 213-14; John Hall, *The History of Broadcasting in New Zealand: 1920-1954*, pp. 4-6; Patrick Day, *The Radio Years: A history of broadcasting in New Zealand Volume One*, pp. 6-33; A.C. Wilson, *Wire & Wireless: A history of telecommunications in New Zealand 1890-1987*, pp. 91-96; Chris Foote, 'Broadcasting in New Zealand and the Role of the Post Office: Part One', *The Mail Coach: Journal of the Postal History Society of New Zealand*, 47, 2, 2010, pp. 66-80; Foote, 'Part Two', *Mail Coach*, 48, 3, 2012, pp. 104-19; Foote, 'Part Three', *Mail Coach*, 48, 5, 2012, pp. 186-95.
10. Hall, *The History of Broadcasting*, p. 2; Day, *The Radio Years*, p. 13; Wilson, *Wire & Wireless*, p. 92; Ian Dougherty, *Ham Shacks, Brass Pounders & Rag Chewers: A history of amateur radio in New Zealand*, pp. 16-18; Foote, 'Broadcasting in New Zealand Part One', pp. 69-73.
11. *NZS*, 1903, p. 28.

NOTES: Chapter 6 Receiving Radio

12. *New Zealand Parliamentary Debates* (*NZPD*), 1903, Vol. 125, pp. 65–67, 288–89.
13. Ibid., 1903, Vol. 125, p. 289.
14. Wilson, *Wire & Wireless*, p. 92; Foote, 'Broadcasting in New Zealand Part One', p. 74.
15. Day, *Radio Years*, p. 27.
16. *NZS*, 1913, pp. 26–70; Day, *Radio Years*, pp. 28, 30; Foote, 'Broadcasting in New Zealand Part One', p. 79.
17. Day, *Radio Years*, p. 30; Foote, 'Broadcasting in New Zealand Part One, pp. 69–73.
18. Dougherty, *Ham Shacks*, pp. 21–22.
19. Robinson, *A History of the Post Office*, pp. 182–83; Day, *Radio Years*, pp. 12–26; Wilson, *Wire & Wireless*, pp. 91–96; Foote, 'Broadcasting in New Zealand Part One', pp. 75–77.
20. Robinson, *A History of the Post Office*, p. 183.
21. Day, *Radio Years*, p. 24.
22. *Appendices to the Journals of the House of Representatives* (*AJHR*), 1914, F–01 p. 11.
23. *EP*, 19 September 1913, p. 8.
24. Ian Dougherty, 'Bell, Francis Wirgman Dillon 1896–1987'; 'Bell, Margaret Brenda 1891–1979', Dictionary of New Zealand Biography, www.teara.govt.nz/en/biographies/4b20/1
25. Susan J. Douglas, *Inventing American Broadcasting 1899–1922*, pp. 187–94.
26. *Poverty Bay Herald*, 21 December 1907, p. 7; *Taranaki Herald*, 31 December 1907, p. 4; *Otago Witness* (*OW*), 1 January 1908, p. 28; *Auckland Star*, 4 January 1908, p. 10; *OW*, 22 January 1908, p. 5; *EP*, 19 December 1908, p. 6; *Taranaki Herald*, 23 December 1908, p. 4; *OW*, 6 January 1909, p. 40.
27. James Belich, *Paradise Reforged: A history of the New Zealanders from the 1880s to the Year 2000*, p. 74.
28. Jock Phillips, *A Man's Country? The image of the Pakeha male: A history*, pp. 132–58.
29. Hall, *History of Broadcasting*, p. 3.
30. *EP*, 11 September, 1908, p. 4; *Feilding Star*, 11 September 1908, p. 3; *Nelson Evening Mail*, 11 September 1908, p. 4; *Hawera & Normanby Star*, 11 September 1908, p. 5.
31. *EP*, 11 September 1908, p. 4; *Marlborough Express*, 17 September 1908, p. 6.
32. *OW*, 16 September 1908, p. 66; *Marlborough Express*, 17 September 1908, p. 6.
33. Dougherty, *Ham Shacks*, p. 18.
34. Section 50 of the 1909 Shipping and Seamen Amendment Act began the process of making wireless compulsory for New Zealand-registered ships. *NZS*, 1909, pp. 339–40. Wireless began to be featured in shipping line advertising from 1909. *Taranaki Herald*, 4 November 1909, p. 1.
35. Day, *Radio Years*, p. 20; Dougherty, *Ham Shacks*, p. 18.
36. *AJHR*, 1910, F-8, pp. 24–25.
37. *Auckland Star*, 8 October 1910; *Poverty Bay Herald*, 13 October 1910, p. 4.
38. Dougherty, *Ham Shacks*, p. 18.

223

39. Day, *Radio Years*, pp. 20–22, 27–30; Dougherty, *Ham Shacks*, pp. 19–23; Foote, 'Broadcasting in New Zealand Part One', pp. 78–79.
40. L.S.S. Spackman, 'When Radio Was Very Young: The story of New Zealand radio in the pre-company days', *N.Z. Radio Times*, 1, 1, 1932, p. 2.
41. Michael Bull, *Sounding Out the City: Personal stereos and the management of everyday life*; Bull, 'Thinking About Sound, Proximity and Distance in Western Experience: The case of Odysseus's walkman', in Veit Erlmann, ed., *Hearing Cultures: Essays on sound, listening and modernity*, pp. 173–91; Bull, *Sound Moves: iPod culture and urban experience*; Heike Weber, 'Head Cocoons: A sensori-social history of earphone use in West Germany 1950–2010', *Senses & Society*, 5, 3, 2010, pp. 339–63; Kate Crawford, 'Four Ways of Listening with an iPhone: From sound and network listening to biometric data and geolocative tracking', in Larissa Hjorth, Jean Burgess and Ingrid Richardson, eds, *Studying Mobile Media: Cultural technologies, mobile communication and the iPhone*, pp. 213–28.
42. Jonathan Sterne, *The Audible Past: Cultural origins of sound reproduction*, pp. 154–77.
43. Ken Collins, *Broadcasting Grave and Gay*, p. 13.
44. Tarpan Sarkar et al., *History of Wireless*, pp. 355–57.
45. *EP*, 27 December 1911, p. 7.
46. Peter Downes, 'Drummond, David Archibald Victor Clive', Dictionary of New Zealand Biography, www.teara.govt.nz/en/biographies/4d20/drummond-david-archibald-victor-clive
47. Barbara Basham, *Portrait From Life: Clive Drummond*, D Series. Id No. 33551, Radio New Zealand Sound Archives.
48. *Progress*, 1, 1, 1905, p. 5.
49. Louis E. Walker, 'Wireless Telegraphy', *Progress*, 2, 3, 1907, pp. 92–94.
50. J. Cowan, *Official Record of the New Zealand International Exhibition of Arts and Industries, Held at Christchurch 1906-7: A descriptive and historical account*, pp. 141–42; Day, *Radio Years*, pp. 16–17, 20–21; Foote, 'Broadcasting in New Zealand Part One', p. 74.
51. G.A.P., 'Details of the Marconi System', *Progress*, 2, 3, 1907, p. 94.
52. *New Zealand Illustrated Magazine*, 1 May 1900, p. 567; *Popular Mechanics*, 7, 7, p. 756; 8, 4, p. 274; *Colonist*, 4 April 1910, p. 4; *Hawera & Normanby Star*, 20 July 1910, p. 3; *Rodney and Otamatea Times*, 1 January 1913, p. 7.
53. Day, *Radio Years*, pp. 16–17.
54. Jock Phillips, 'Exhibiting Ourselves: The exhibition and national identity', in J.M. Thomson, ed., *Farewell Colonialism: The New Zealand Exhibition Christchurch 1906-07*, pp. 17–26.
55. *Taranaki Herald*, 17 September 1897, p. 2; *Bay of Plenty Times*, 11 October 1897, p. 2; *EP*, 16 October 1897, p. 5.
56. Wilson, *Wire & Wireless*, pp. 47–50.
57. Caroline Daley, 'Modernity, Consumption and Leisure', in G. Byrnes, ed., *The New Oxford History of New Zealand*, pp. 423–28.
58. Day, *Radio Years*, pp. 15–19; Wilson, *Wire & Wireless*, pp. 91–96.

59. Susan Douglas, *Listening In: Radio and the American imagination*, pp. 41–42.
60. R.S. Ellwood, *Islands of the Dawn: The story of alternative spirituality in New Zealand*, pp. 27–57.
61. *Wanganui Herald*, 2 June 1903, p. 6.
62. Joseph Taylor, 'Principles of Astrology', *Nelson Evening Mail*, 30 October 1906, p. 1.
63. *OW*, 19 August 1908, p. 14.
64. Ibid.
65. Jeffrey Sconce, *Haunted Media: Electronic presence from telegraphy to television*, pp. 66–75.

Chapter 7 Military Radio

1. Desmond Ball, Cliff Lord and Meredith Thatcher, *Invaluable Service: The secret history of New Zealand's signals intelligence during two world wars*, pp. 3–6.
2. Peter Downes, 'Drummond, David Archibald Victor Clive', Dictionary of New Zealand Biography, www.teara.govt.nz/en/biographies/4d20/drummond-david-archibald-victor-clive
3. Matthew Tonks, 'The New Zealand Expeditionary Force Sets Forth', from New Zealand WW100, http://ww100.govt.nz/the-new-zealand-expeditionary-force-sets-forth
4. John Hall, *The History of Broadcasting in New Zealand 1920–1954*, p. 4; Patrick Day, *The Radio Years: A history of broadcasting in New Zealand Volume One*, p. 32; Ball, Lord and Thatcher, *Invaluable Service*, p. 7.
5. Hall, *History of Broadcasting*, p. 4; Day, *The Radio Years*, p. 31.
6. Stephen Smith, *The Samoa (N.Z.) Expeditionary Force 1914–1915*, p. 7; Ian McGibbon, ed., *The Oxford Companion to New Zealand Military History*, pp. 475–76.
7. Malama Meleisea, Penelope Schoeffel Meleisea, Gatoloai Peseta S. Sio and I`iga Suafole, 'German Samoa 1900–1914', in Malama Meleisea and Penelope Schoeffel Meleisea (eds), *Lagaga: A short history of Western Samoa*, Suva, 1987, p. 108.
8. Stephen John Smith, 'The Seizure and Occupation of Samoa', in H.T.B. Drew, *The War Effort of New Zealand. Official History of New Zealand's Effort in the Great War*, pp. 23–41; Stephen John Smith, *The Samoa (N.Z.) Expeditionary Force 1914–1915*; Mary Boyd, 'The Military Administration of Western Samoa, 1914–1919', *New Zealand Journal of History*, 2, 2, 1968, pp. 148–64; Ian McGibbon, *The Path to Gallipoli: Defending New Zealand 1840–1915*; Ian McGibbon, 'The Shaping of New Zealand's War Effort, August–October 1914', in John Crawford and Ian McGibbon, eds, *New Zealand's Great War: New Zealand, the Allies & the First World War*, pp. 49–68; Ian McGibbon, 'New Zealand's First Conquest', *New Zealand International Review*, 39, 6, 2014, p. 18.
9. McGibbon, 'New Zealand's First Conquest', p. 18.
10. Ball, Lord and Thatcher, *Invaluable Service*, p. 9.
11. Smith, *The Samoa (N.Z.) Expeditionary Force 1914–1915*, pp. 112–19.
12. Peter Hoar, 'A Qualitative Content Analysis of the New Zealand Troopship Publications 1914–1920', MLIS thesis, Victoria University of Wellington, 2001;

Joanna Condon, '"Mainly About Us": Identity and marginality in the troop magazines and newspapers of the New Zealand Expeditionary Force 1914–1919', *NZ Library and Information Management Journal*, 52, 3, 2012, www.lianza.org.nz/nz-library-and-information-management-journal-vol-52-no-3
13. Smith, *The Samoa (N.Z.) Expeditionary Force 1914–1915*, pp. 105–11.
14. *The Pull Thro'*, 1, 2, 1914, p. 2.
15. Ibid., p. 10.
16. *New Zealand Herald (NZH)*, 6 March 1915, p. 10; *Oamaru Mail*, 1 April 1915, p. 5; *Sun*, 15 April 1915, p. 3.
17. Day, *The Radio Years*, p. 31.
18. *NZH*, 31 January 1918, p. 6.
19. *Auckland Star*, 30 January 1918, p. 2.
20. *Southland Times*, 22 May 1918, p. 4.
21. *Sun*, 28 June 1918, p. 3.
22. *Southland Times*, 13 October 1920, p. 4.
23. 'Atom', 'Wireless', *The Kia Ora Coo-Ee: The magazine for the Anzacs in the Middle East, 1918*, Sydney, Facsimile, 1981, Second Series, p. 2; H.T.B. Drew, *The War Effort of New Zealand*, pp. XXI–XXII; C.G. Powles, H.G. Alexander and H.A. Lockington, *Official History of the New Zealand Engineers During the Great War 1914–1919*, pp. 299–303; Day, *The Radio Years*, pp. 30–33; A.C. Wilson, *Wire & Wireless: A history of telecommunications in New Zealand 1890–1987*, pp. 110–13.
24. Briton Cooper Busch, *Britain, India, and the Arabs 1914–1921*. Berkeley, 1971; Ron Wilcox, *Battles on the Tigris: The Mesopotamian Campaigns of the First World War*, Barnsley, 2006; A.J. Barker, *The First Iraq War, 1914–1918: Britain's Mesopotamian Campaign*, New York, 2009; Charles Townsend, *When God Made Hell: The British invasion of Mesopotamia and the Creation of Iraq 1914–1921*, London, 2010; James Barr, *A Line in the Sand: The Anglo-French struggle for the Middle East 1914–1948*, New York, 2012; Ian Rutledge, *Enemy on the Euphrates: The battle for Iraq 1914–1921*, London, 2014; Eugene Rogan, *The Fall of the Ottomans: The great war in the Middle East 1914–1920*, London, 2015; N.S. Nash, *The Betrayal of an Army: Mesopotamia 1914–1916*, Barnsley, 2016.
25. Glyn Harper, *Johnny Enzed: The New Zealand soldier in the First World War 1914–1918*, Auckland, 2015, p. 515.
26. *EP*, 24 February 1916, p. 3
27. C.G. Powles, H.G. Alexander and H.A. Lockington, 'N.Z. Wireless Troop in Mesopotamia and Persia', in N. Annabell, ed., *Official History of the New Zealand Engineers during the Great War, 1914–1919: A record of the work carried out by the field companies, field troops, Signal Troop and Wireless Troop, during the operations in Samoa (1914–15); Egypt, Gallipoli, Sinai and Palestine (1914–1918); France, Belgium and Germany (1916–1919); and Mesopotamia (1916–1918)*, pp. 298–99.
28. *Willochra Tatler*, 1, 1, 1916, cover.
29. Ibid., p. 17.
30. Ibid., p. 10.
31. Ibid., p. 7.

32. *NZH*, 28 November 1916, p. 9.
33. H.T.B. Drew, *The War Effort of New Zealand*, p. xxii; Powles, Alexander and Lockington, 'N.Z. Wireless Troop', pp. 299–300; A.C. Wilson, *Wire & Wireless: A history of telecommunications in New Zealand 1890–1987*, pp. 110-11; Harper, *Johnny Enzed*, p. 515.
34. Harper, *Johnny Enzed*, p. 515.
35. Auckland War Memorial Museum Online Cenotaph, 'Richard Croucher', http://aucklandmuseum.com/war-memorial/online-cenotaph/record/C65128. Croucher died in Palestine in 1919 after breaking his back while swimming.
36. *Marlborough Express*, 25 October 1916, p. 6.
37. *Sun*, 5 October 1916, p. 7; *Hawera & Normanby Star*, 11 October 1916, p. 4; *Grey River Argus*, 14 October 1916, p. 2.
38. Wilson, *Wire & Wireless*, p. 111.
39. *Dominion*, 22 July 1916, p. 5; *Wairarapa Daily Times*, 24 August 1916, p. 2; 5 December 1916, p. 5; 16 October 1917, p. 5.
40. Charles Bean, *The Australian Imperial Force in France during the Main German Offensive, 1918*, Sydney, 8th edn, 1941, pp. 711–12.
41. Keast Burke, ed., *With Horse and Morse in Mesopotamia: The story of Anzacs in Asia*, Sydney, 1927, p. 37.
42. Ibid.
43. *EP*, 6 August 1917, p. 9; *NZH*, 11 August 1917, p. 8; Bean, *The Australian Imperial Force in France during the Main German Offensive, 1918*, pp. 713–14.
44. Ibid., pp. 712–13.
45. G.S. Richardson, 'Education in the New Zealand Expeditionary Force', in Drew, *The War Effort of New Zealand*, pp. 220–32.
46. Hall, *History of Broadcasting*, p. 4.
47. Ibid.
48. Dougherty, *Ham Shacks*, p. 22.
49. *Auckland Star*, 4 June 1928, p. 3; *NZH*, 5 June 1928, p. 10; *New Zealand Truth*, 21 June 1928, p. 6; *Ellesmere Guardian*, 22 June 1928, p. 2; Jane Latchem, 'A Generation of Promise: The 1908 Junior National Scholarship Candidates – education, occupation, and the First World War', *Journal of New Zealand Studies*, 18, 2014, pp. 74–75.
50. Clive Drummond, 'Early Radio in New Zealand: Interesting and entertaining reminiscences', *N.Z. Radio Record*, 2, 7, 1928, p. 4.
51. Downes, 'Drummond, David Archibald Victor Clive 1890–1978'.
52. Peter Downes and Peter Harcourt, *Voices in the Air: Radio broadcasting in New Zealand, a documentary*, pp. 7–20; Hall, *The History of Broadcasting*, pp. 7–16; Day, *The Radio Years*, pp. 38–48.
53. Day, *The Radio Years*, pp. 37–38.
54. L.S.S. Spackman, 'When Radio Was Very Young: The story of New Zealand radio in the pre-company days', *The N.Z. Radio Times*, 1, 1, 1932, p. 3.
55. Day, *The Radio Years*, pp. 39–40.
56. *ODT*, 21 December 1921, p. 7.
57. Day, *The Radio Years*, p. 48.

58. *Press*, 6 January 1923, p. 9.
59. *Auckland Star*, 1 June 1923, p. 11.
60. *NZH*, 12 September 1923, p. 6; *ODT*, 21 September 1923, p. 2; *Bay of Plenty Times*, 22 September 1923, p. 4.
61. Drummond, 'Early Radio', p. 5.

Chapter 8 Hearing Radio

1. *Dominion*, 30 January 1930, p. 4; *Evening Post* (*EP*), 30 January 1930, p. 13, *N.Z. Radio Record* (*NZRR*), 31 January 1930, pp. 8, 12; 7 February 1930, p. 12.
2. Margaret Willis, *Aunt Gwen of 2YA*, pp. 41–43.
3. Gwen made a recording of a version of her show in 1930 when she was in Sydney: Aunt Gwen of 2YA with Uncle Bruce, Columbia No. X39, Sound Archives, Ngā Taonga Kōrero, 27900.
4. Willis, *Aunt Gwen*, pp. 37–39.
5. Ken G. Collins, *Broadcasting Grave and Gay*, p. 30.
6. Gerald Bloomfield, *New Zealand: A handbook of historical statistics*, p. 261.
7. Willis, *Aunt Gwen*, p. 42.
8. *Otago Daily Times* (*ODT*), 21 November 1930, p. 5; Bloomfield, *A Handbook of Historical Statistics*, p. 261.
9. Patrick Day, *The Radio Years: A history of broadcasting in New Zealand Volume One*, p. 321.
10. Downes and Harcourt, *Voices in the Air*, pp. 68–114; Hall, *History of Broadcasting*, pp. 17–25; Day, *The Radio Years*, pp. 63–248. For Māori involvement in radio, see Claudia Lemke, 'Maori Involvement in Sound Recording and Broadcasting 1919 to 1958', pp. 89–139. For radio hardware, see John W. Stokes, *The Golden Age of Radio in the Home*. For a personal account by a broadcaster, see Collins, *Broadcasting Grave and Gay*, pp. 24–82. For amateur operators (Ham radio), see Ian Dougherty, *Ham Shacks, Brass Pounders & Rag Chewers: A history of amateur radio in New Zealand*, pp. 37–98.
11. Stokes, *The Golden Age of Radio in the Home*, pp. 16–30.
12. Benedict Anderson, *Imagined Communities: Reflections on the origins and spread of nationalism*, pp. 6–7.
13. The 2010 film about the 1937 coronation of King George VI, *The King's Speech*, may have contained many inaccuracies but it illustrated very clearly the importance of radio for national and imperial unity.
14. Rudolf Arnheim, *Radio*, pp. 232–33.
15. Michele Hilmes, *Radio Voices: American broadcasting 1922–1952*, pp. 11–23; Carolyn Birdsall, *Nazi Soundscapes: Sound, technology and urban space in Germany, 1933–1945*, pp. 103–39; Keith Somerville, *Radio Propaganda and the Broadcasting of Hatred*, pp. 33–54.
16. Susan Douglas, *Listening In: Radio and the American imagination*, p. 24.
17. Ibid., pp. 33–34.
18. Karin Bijsterveld and José van Dijck, 'Introduction', in Karin Bijsterveld and José van Dijck, eds, *Sound Souvenirs: Audio technologies, memory and cultural practices*, pp. 14–16.

19. Arnheim, *Radio*, pp. 258–68; Shaun Vancour, 'Arnheim on the Radio: *Materialtheorie* and beyond', in Scott Higgins, ed., *Arnheim for Film and Media Studies*, pp. 177–94.
20. Shaun Moores, '"The Box on the Dresser": Memories of early radio and everyday life', *Media, Culture and Society*, 10, 1, 1988, pp. 31–33; Day, *The Radio Years*, pp. 94–96; Shelley Nickles, 'Object Lessons: Household appliance design and the American middle class, 1920–1960', pp. 158–218; Artemis Yagou, 'Shaping Technology for Everyday Use: The case of the radio set design', *The Design Journal*, 5, 1, 2002, pp. 2–13.
21. Stokes, *The Golden Age of Radio in the Home*, p. 28.
22. Barbara Anderson, *Getting There: An autobiography*, p. 87.
23. Lorraine Russell, *Family Hold Back: A 1930s childhood*, p. 79.
24. Ibid.
25. *New Zealand Woman's Weekly*, 20 February 1936, p. 29.
26. *New Zealand Woman's Weekly*, 2 April 1936, p. 31.
27. Bloomfield, *Historical Statistics*, p. 261.
28. *NZRR*, 3 August 1928, pp. 4–5.
29. *ODT*, 4 June 1926, p. 14; *New Zealand Herald*, 6 September 1926, p. 14; *Auckland Star*, 8 September 1926, p. 9; *Press*, 2 October 1926, p. 10; *EP*, 30 October 1926, p. 11.
30. Tony Currie, *The Radio Times Story*.
31. *NZRR*, 6 April 1928, p. 3.
32. The following account is based on *NZRR*, 9 September 1927, pp. 8–9. See *NZRR* passim for further examples.
33. George Clark, 'Mr George Clark' in Ken Clover, ed., *The People of the Plains: Some early memories of those who have lived in or been associated with the Hauraki Plains Volume 1*, p. 169.
34. Dawn Sheppard, 'Seventy Years of Radio Listening', in Jim Sullivan, ed., *As I Remember: Stories from 'Sounds Historical' Volume 2*, pp. 78–79.
35. Patrick Day, 'Broadcasting the Centennial', in William Renwick, ed., *Creating a National Spirit: Celebrating New Zealand's Centennial*, Wellington, 2004, p. 81.
36. Moores, '"The Box on the Dresser"', pp. 35–39.
37. Paddy Scannell, *A Social History of British Broadcasting: Serving the nation 1922–1939*, pp. 371–72.
38. Sian Nicholas, 'The Reithian Legacy and Contemporary Public Service Ethos', in M. Conboy and J. Steel, eds, *The Routledge Companion to British Media History*, pp. 323–33.
39. A.R. Harris, 'The Policy of Service behind Radio in New Zealand', *NZRR*, 3 August 1928, p. 2.
40. *NZRR*, 6 April 1928, p. 3.
41. Collins, *Broadcasting Grave and Gay*, pp. 25–26.
42. For an account of the 1925 regulations, see Day, *The Radio Years*, pp. 60–61.
43. Hall, *History of Broadcasting*, p. 23.
44. *NZRR*, 2 November 1928, p. 24.
45. *NZRR*, 5 October 1928, p. 31.

46. *NZRR*, 19 October 1928, p. 25.
47. *NZRR*, 23 September 1927, p. 4.
48. Examples include: *NZRR*, 28 October 1927, pp. 4–7; Mary Jordan, 'Radio Spreads the Love of Good Music', ibid., 23 March 1928, p. 1 ; 'Brilliant Orchestra Engaged for 2YA', ibid., 27 April 1928, p. 5; 'What Is an Opera', ibid., 6 July 1928, p. 3; 'Grand Opera Series From All Stations', ibid., 7 December, 1928, pp. 1–2.
49. Day, *The Radio Years*, pp. 88–90.
50. *EP*, 25 September 1933, p. 8.
51. 'Mrs Betty Mules (nee McClean)', in Clover, ed., *The People of the Plains Volume 3*, p. 621.
52. Patricia Ridding, 'Living with 2ZB in the Thirties', in Sullivan, ed., *As I Remember*, p. 83.
53. *NZRR*, 5 August 1927, p. 4.
54. Emily Thompson, *The Soundscape of Modernity: Architectural acoustics and the culture of listening in America 1900–1933*, pp. 149–52; Karin Bijestereald, *Mechanical Sound: Technology, culture, and public problems of noise in the twentieth century*, pp. 159–92.
55. D.G.C., 'Less Noise, Please', *New Zealand Woman's Weekly*, 15 October 1936, p. 11.
56. *Auckland Star*, 7 December 1926, p. 16; 18 January 1930, p. 8; *EP*, 11 July 1935, p. 23.
57. Ibid., 28 March, 1935, p. 8.

Chapter 9 Organising Radio

1. *Evening Post* (*EP*), 9 March 1932 p. 9; 8 September 1932, p. 19; *Auckland Star*, 9 September 1932, p. 12; Ian K. MacKay, *Broadcasting in New Zealand*, pp. 43–46; Ken Collins, *Broadcasting Grave and Gay*, pp. 52–53; Patrick Day, *The Radio Years: A history of broadcasting in New Zealand Volume One*, pp. 163–64.
2. Gerald Bloomfield, *New Zealand: A handbook of historical statistics*, p. 261.
3. *Press*, 7 September 1932, p. 10; *EP*, 8 September 1932, p. 19.
4. *Auckland Star*, 9 March 1932, p. 8.
5. New Zealand Broadcasting Service, Plebiscite of Listeners, 1932, AAFK, 890, 22c, Archives New Zealand (ANZ), Wellington.
6. H.N. Bullock to Broadcasting Board, in ibid.
7. *EP*, 8 September 1932, p. 19.
8. Ibid.
9. G. Morley to Broadcasting Board, New Zealand Broadcasting Service, Plebiscite of Listeners.
10. 'Listener in' to Broadcasting Board, in ibid.
11. D. Murray to Broadcasting Board, in ibid.
12. *Auckland Star*, 14 January 1932, p. 6.
13. Cecelia Tichi, *Electronic Hearth: Creating an American television culture*, pp. 42–61.
14. *EP*, 10 March 1932, p. 10.
15. *EP*, 11 March 1932, p. 6.
16. *EP*, 12 March 1932, p. 10.

17. Day, *Radio Years*, pp. 80–83.
18. David Oswell, 'Early Children's Broadcasting in Britain: Programming for a liberal democracy', *Historical Journal of Film, Radio and Television*, 18, 3, 1998, pp. 379–80; Bridget Griffen-Foley, 'Modernity, Intimacy and Early Australian Commercial Radio', in Joy Damousi and Desley Deacon, eds, *Talking and Listening in the Age of Modernity: Essays on the history of sound*, pp. 123–24.
19. Lesley Johnson, *The Unseen Voice: A cultural study of early Australian radio*, p. 23.
20. Asa Briggs, *The Birth of Broadcasting: The history of broadcasting in the United Kingdom Volume I*, pp. 258–62.
21. Day, *Radio Years*, pp. 81–82.
22. Christchurch City Libraries, *Ernest John Bell: 1885?–1971*, n.d., http://christchurchcitylibraries.com/Heritage/People/B/BellErnestJ/; *Remembering Jack and Edna*, n.d., http://library150.com/Articles/ErnestBell/
23. *New Zealand Radio Record* (*NZRR*), 27 July 1928, p. 3.
24. Fred Price to Broadcasting Board, New Zealand Broadcasting Service, Plebiscite of Listeners.
25. *NZRR*, 2, 3, 3 August 1928, pp. 49–51.
26. Donald Webster, 'An Old Man's Radio Ramblings', in Sullivan, ed., *As I Remember Vol. 2*, p. 31.
27. Mary Matthews, 'Our Radio', *Sounds Historical*, 2 March 2008, www.radionz.co.nz/national/programmes/soundshistorical/20080302
28. Falconer to 2YA, 11 September 1928, Gwen Shepherd, Ms Papers 7133-1, Alexander Turnbull Library, Wellington. A similar letter is found in 'A Mother's Request', *NZRR*, 11 May 1928, p. 16.
29. Hall, *History of Broadcasting*, pp. 36–38; Day, *Radio Years*, pp. 79–81.
30. *New Zealand Truth*, 25 July 1929, p. 12.
31. *New Zealand Truth*, 12 September 1929, p. 12.
32. *EP*, 21 June 1928, p. 19; *Auckland Star*, 6 June 1930, p. 13; *EP*, 8 June 1935, p. 26.
33. Katie Day Good, 'Radio's Forgotten Visuals', *Journal of Radio and Audio Media*, 23, 2, 2016, pp. 364–68.
34. *Auckland Star*, 13 June 1930, p. 13.
35. *Auckland Star*, 25 July 1930, p.18.
36. *Auckland Star*, 1 August 1930, p. 15.
37. Day, *The Radio Years*, pp. 113–18.
38. Ibid., pp. 167–70.
39. Patrick Day, 'American Popular Culture and New Zealand Broadcasting: The reception of early radio serials', *Journal of Popular Culture*, 30, 1, 1996, p. 209.
40. 'John Hill' in Ken Clover, ed., *Hauraki Memories: Some treasured memories from the wider Hauraki area*, p. 55.
41. Paul Byrnes, *Curator's Notes: Dad and Dave from Snake Gully, Episode 1 (1937)*, http://aso.gov.au/titles/radio/dad-and-dave-snake-gully-ep-1/notes/; Albert Moran and Chris Keating, *The A to Z of Australian Radio and Television*, p. 124.
42. Gordon McLauchlan, *A History of New Zealand Humour*, Auckland, 1989, pp. 33–39, 197–88; Lydia Wevers, 'The Short Story', in Terry Sturm, ed., *The

Oxford History of New Zealand Literature in English, pp. 264–65.
43. Barbara Anderson, *Getting There: An autobiography*, p. 88.
44. 'Mrs Betty Mules (nee McLean)', in Clover, ed., *The People of the Plains Volume 3*, p. 621.
45. Day, *The Radio Years*, pp. 123–24.
46. Ibid., p. 124.
47. Collins, *Broadcasting Grave and Gay*, p. 32.
48. Donna Beatson, 'A Genealogy of Maori Broadcasting: The development of Maori radio', *Continuum: Journal of Cultural and Media Studies*, 10, 1, 1996, pp. 76–93; Ian Stuart, 'The Construction of National Maori Identity by Maori Media', *Pacific Journalism Review*, 9, 2003, pp. 45–58.
49. Day, *Radio Years*, p. 130; Simon J. Potter, *Broadcasting Empire: The BBC and the British world 1922–1970*, p. 38.
50. *EP*, 12 November 1927, p. 10; *Auckland Star*, 14 November 1927, p. 8; *Ellesmere Guardian*, 18 November 1927, p. 6.
51. *EP*, 23 April 1936, p. 23.
52. Russell, *Family Hold Back*, p. 80.
53. Paddy Scannell, *A Social History of British Broadcasting: Serving the nation 1922–1939*, p. 284.
54. Potter, *Broadcasting Empire*, p. 94.
55. *Press*, 15 May 1937, p. 19.
56. Peter Gibbons, 'The Far Side of the Search for Identity: Reconsidering New Zealand history', *New Zealand Journal of History*, 37, 1, 2003, p. 44.

Chapter 10 Living Pictures

1. *Auckland Star (AS)* 29 November 1895, p. 2; 30 November 1895, p. 8; *New Zealand Herald (NZH)*, 2 December 1895, p. 4; *New Zealand Observer*, 21 December 1895, p. 9; Clive Sowry, 'Edison's Kinetoscope', *The Big Picture*, 8, 1996, pp. 11–12; Clive Sowry, 'Whitehouse, Alfred Henry 1856–1929', Dictionary of New Zealand Biography, www.teara.govt.nz/en/biographies/2w16/1; Chris Pugsley, 'The Magic of Moving Pictures: Film making 1895–1918', in Diane Pivac, Frank Stark and Lawrence McDonald, eds, *New Zealand Film: An illustrated history*, pp. 29–31.
2. W.K.L. Dickson and Antonia Dickson, *History of the Kinetograph, Kinetoscope and Kineto-Phonograph*; Charles Musser, *The Emergence of Cinema: The American screen to 1907*, pp. 62–89; Ray Phillips, *Edison's Kinetoscope and its Films: A history to 1896*, pp. 27–36; Paul C. Spehr, 'Movies and the Kinetoscope 1890–1895', in André Gaudreault, ed., *American Cinema 1880–1909: Themes and variations*, pp. 22–44.
3. Sowry, 'Edison's Kinetoscope', p. 11.
4. Rosalind Rogoff, 'Edison's Dream: A brief history of the kinetophone', *Cinema Journal*, 15, 2, 1976, pp. 58–68; Musser, *The Emergence of Cinema*, pp. 87–88; Phillips, *Edison's Kinetoscope*, pp. 75–89; Rick Altman, *Silent Film Sound*, pp. 78–83. This is a quite separate device from Edison's short-lived 1913 kinetophone. See Altman, *Silent Film Sound*, pp. 175–78.

5. *AS*, 16 January 1896, p. 8; *New Zealand Observer*, 25 January 1896, p. 16.
6. *AS*, 16 January 1896, p. 8.
7. *Wanganui Chronicle*, 19 April 1894, p. 2; *Evening Post* (*EP*), 18 May 1894, p. 4; *Nelson Evening Mail*, 3 July 1894, p. 2.
8. *Star*, 19 May 1894, p. 1.
9. Clive Sowry, 'Non-Fiction Films: Between the wars', in Pivac et al., eds, *New Zealand Film*, p. 94.
10. Altman, *Silent Film Sound*, pp. 193–201.
11. Pivac, Stark and McDonald, eds, *New Zealand Film*, pp. 30, 74–77, 94–97.
12. The following anthologies provide a good cross-section of the variety of approaches involved in the burgeoning field of Sound Studies: Michael Bull and Les Back, eds, *The Auditory Culture Reader*; Jim Drobnick, ed., *Aural Cultures*; Veit Erlmann, ed., *Hearing Cultures: Essays on sound, listening and modernity*; Mark M. Smith, ed., *Hearing History: A reader*; David Suisman and Susan Strasser, eds, *Sound in the Age of Mechanical Reproduction* ; Jonathan Sterne, ed., *The Sound Studies Reader*; Trevor Pinch and Karin Bijstervald, eds, *The Oxford Handbook of Sound Studies*.
13. Philip Brophy, 'Where Sound Is: Locating the absent aural in film theory', in James Donald and Michel Renov, eds, *The Sage Handbook of Film Studies*, Los Angeles, 2008, pp. 425–35.
14. For a comprehensive survey of such techniques and technologies, see Altman, *Silent Film Sound*.
15. Walter Benjamin, 'The Work of Art in the Age of Mechanical Reproduction', in Hannah Arendt, ed., *Illuminations*, pp. 217–51; Theodor W. Adorno and Max Horkheimer, *Dialectic of Enlightenment*, pp. 120–67; James Lastra, *Sound Technology and the American Cinema: Perception, representation, modernity*, p. 4.
16. Clive Sowry, 'MacMahon's Cinematographe', *The Big Picture*, 10, 1996, p. 22.
17. *Otago Witness* (*OW*), 5 November 1896, p. 9; *Timaru Herald*, 17 December 1896, p. 3; *Poverty Bay Herald*, 27 February 1897, p. 2.
18. *West Coast Times*, 23 April 1900, p. 4.
19. *Wanganui Herald*, 25 May 1900, p. 2.
20. *OW*, 31 May 1900, p. 47.
21. *Wanganui Herald*, 19 July 1900, p. 2; *Taranaki Herald* (*TH*), 11 August 1900, p. 2.
22. *TH*, 10 August 1900, p. 2.
23. *Hawera & Normanby Star* (*HNS*), 1 August 1900, p. 2.
24. *West Coast Times*, 23 April 1900, p. 3; *Wanganui Herald*, 20 July 1900, p. 3.
25. *TH*, 10 August 1900, p. 2.
26. *OW*, 30 August 1900, p. 55; *TH*, 19 October 1900, p. 2.
27. *Poverty Bay Herald*, 1 November 1900, p. 2.
28. *New Zealand Free Lance*, 1 December 1900, p. 8.
29. Musser, *Emergence of Cinema*, pp. 337–69; Tom Gunning, 'Now You See It, Now You Don't: The temporality of the cinema of attractions', in Lee Grieveson and Peter Krämer, eds, *The Silent Cinema Reader*, p. 43; Richard Abel, 'The

Cinema of Attractions in France 1896–1904', in Grieveson and Krämer, eds, *Silent Cinema Reader*, pp. 63–75; Charles Musser, 'Moving Towards Fictional Narratives: Story films become the dominant product 1903–1904', in Grieveson and Krämer, *Silent Cinema Reader*, pp. 87–102.
30. *HNS*, 31 October 1910, p. 4; *Grey River Argus*, 28 November 1910, p. 5.
31. *EP*, 12 April 1910, p. 2; *HNS*, 1 November 1910, p. 5.
32. *EP*, 8 April 1910, p. 2.
33. *EP*, 12 April 1910, p. 2.
34. *Grey River Argus*, 18 October 1910, p. 10.
35. *EP*, 8 July 1940, p. 9.
36. Ibid., 23 June 1915, p. 3; *Ashburton Guardian*, 25 October 1916, p. 3; *Akaroa Mail and Banks Peninsula Advertiser*, 19 December 1916, p. 2.
37. *EP*, 2 November 1914, p. 3.
38. *Grey River Argus*, 17 June 1918, p. 4; Mary Barr and Jim Barr, 'The Kid From Timaru', *The New Zealand Film Archive*, n.d., www.filmarchive.org.nz/tracking-shots/close-ups/KidFromTimaru.html
39. Bruce Mason's *End of the Golden Weather* (1959) is a famous later example.
40. Jim Sullivan, *Canterbury Voices*, pp. 138–43.
41. *Grey River Argus*, 8 June 1918, p. 4.
42. William Main, '"The Lanthorn that Shews Tricks": The Magic Lantern in Nineteenth-Century New Zealand', *Turnbull Library Record*, 27, 1994, pp. 45–53; Elizabeth Hartrick, 'Consuming Illusions: The magic lantern in Australia and Aotearoa/New Zealand 1850–1910'; Altman, *Silent Film Sound*, pp. 55–72.
43. Cyril R. Bradwell, *Fight the Good Fight: The story of the Salvation Army in New Zealand 1883–1983*, pp. 71–72; Simon Price, *New Zealand's First Talkies: Early film-making in Otago and Southland 1896–1939*, pp. 9–10; Churchman, ed., *Celluloid Dreams*, pp. 7, 10.
44. *West Coast Times*, 11 November 1901, p. 4; *EP*, 29 July 1903, p. 3; *Grey River Argus*, 9 August 1907, p. 3.
45. *Wanganui Herald*, 25 May 1901, p. 3; *Tuapeka Times*, 9 June 1909, p. 3.
46. Joseph L. Anderson, 'Spoken Silents in the Japanese Cinema: Essay on the necessity of Katsuben', *Journal of Film and Video*, 40, 1, 1988, pp. 13–33; Jeffrey Dym, 'Benshi and the Introduction of Motion Pictures to Japan', *Monumenta Nipponica*, 55, 4, 2000, pp. 509–36; Hideaki Fujiki, 'Benshi as Stars: The irony of the popularity and respectability of voice performers in Japanese cinema', *Cinema Journal*, 45, 2, 2006, pp. 68–84; Kuei-Fen Chiu, 'The Question of Translation in Taiwanese Colonial Cinematic Space', *Journal of Asian Studies*, 70, 1, 2011, pp. 77–97; Roald Maliangkay, 'The Power of Representation: Korean movie narrators and authority', *Journal of Korean Studies*, 16, 2, 2011, pp. 213–29; Akiko Mizoguchi, 'Gender and the Art of Benshi: In dialogue with Midori Sawato', *Camera Obscura*, 26, 78, 2011, pp. 154–66.
47. Bruce W. Hayward and Selwyn P. Hayward, *Cinemas of Auckland 1896–1979*, p. 8; Richard Abel and Rick Altman, eds, *The Sounds of Early Cinema*, pp. 133–43; Altman, *Silent Film Sound*, pp. 55–72.
48. Jean-Paul Sartre, *The Words*, p. 123.

49. David Bordwell, Janet Staiger and Kristin Thompson, *The Classical Hollywood Cinema: Film style & mode of production to 1960*, pp. 98–199, 497–517.
50. Jinty Rorke, 'A.N. Whitehouse: An early film pioneer', *Historical Review: Bay of Plenty Journal of History*, 32, 1, 1994, pp. 17–23; Sowry, 'Edison's Kinetoscope', pp. 11–12; Sowry, 'Whitehouse, Alfred Henry 1856–1929'.
51. *New Zealand Observer and Free Lance*, 21 December 1895, p. 9; Sowry, 'Edison's Kinetoscope', p. 11.
52. Churchman, ed., *Celluloid Dreams*, p. 49; Dennis, 'A Time Line', p. 189.
53. Neither Churchman nor Dennis and Bieringa mention Whitehouse's use of recorded sound throughout his career.
54. Rorke, 'A.N. Whitehouse', p. 22.
55. *Timaru Herald*, 24 April 1899, p. 2; *Grey River Argus*, 15 December 1899, p. 4; Rorke, 'A.N. Whitehouse', p. 19.
56. *AS*, 27 January 1897, p. 8; Clive Sowry, 'MacMahon's Cinematographe', pp. 22–23. For the importance of Sandow in New Zealand, see Caroline Daley, *Leisure & Pleasure: Reshaping and Revealing the New Zealand Body 1900–1960*, pp. 1–82.
57. *OW*, 26 November 1896, p. 39; *Grey River Argus*, 28 December 1896, p. 2
58. *Timaru Herald*, 18 December 1896, p. 2.
59. *NZH*, 5 August 1897, p. 5.
60. *NZH*, 7 September 1897, p. 6.
61. *Poverty Bay Herald*, 22 April 1898, p. 2.
62. *AS*, 22 June 1900, p. 8.
63. Maurice Hurst, *Music and the Stage in New Zealand: A century of entertainment 1840–1943*, pp. 29, 68; John Mansfield Thomson, *The Oxford History of New Zealand Music*, p. 134.
64. *EP*, 31 October 1900, p. 5; *Tuapeka Times*, 28 November 1900, p. 3.
65. *Bay of Plenty Times*, 30 November 1904, p. 2.
66. *TH*, 7 July 1906, p. 7; 13 July 1906, p. 7.
67. *AS*, 17 April 1935, p. 8.
68. *Wanganui Chronicle*, 26 June 1906, p. 4; 25 June 1906, p. 5.
69. *New Zealand Observer* (*NZO*), 30 June 1906, p. 4.
70. Richard Brown and Barry Anthony, *A Victorian Film Enterprise: The history of the British Mutoscope and Biograph Company 1897–1915*.
71. *Patea Mail*, 13 July 1906, p. 2.
72. *Wanganui Herald*, 8 August 1906, p. 3.
73. *Grey River Argus*, 23 October 1906, p. 3; *HNS*, 19 November 1906, p. 7.
74. *Tuapeka Times*, 22 September 1906, p. 3.
75. *New Zealand Free Lance*, 17 March 1909, p. 14.
76. *NZO*, 10 April 1909, p. 6; *New Zealand Free Lance*, 8 May 1909, p. 9; *OW*, 26 May 1909, p. 7.
77. *EP*, 25 March 1909, p. 2; *NZO*, 1 May 1909, p. 6.
78. Ingham, *Everyone's Gone to the Movies*, pp. 5–6; Harrison, 'The Motion Picture Industry in New Zealand', pp. 37–93; Knewstubb Theatres, *Cinemas: Dunedin and districts 1897–1974*, pp. 2–4, 6–8, 13; Hayward and Hayward, *Cinemas of*

Auckland, pp. 6–15; Nerida Elliott, 'Anzac, Hollywood and Home: Cinemas and film-going in Auckland 1909-1939', pp. 23–62; Churchman, ed., *Celluloid Dreams*, pp. 9–12; Froude, *Where To Go on Saturday Night*, pp. 7–8.
79. Douglas Lawson, 'At School in Napier in the 1920s', in Sullivan, ed., *As I Remember*, p. 129.
80. Squire Speedy, *The Picturedrome Fun Merchant: Anecdotes of the life and times of the Picturedrome cinema and dancehall at Milford during the era of L.L. Speedy, 1922-1938*, Auckland, 1992, p. 18.
81. Eileen Mathieson, 'The Gaiety Picture Theatre: Paeroa', *Ohinemuri Regional History Journal*, 32, 1988, p. 48.
82. *Bay of Plenty Times*, 26 July 1907, p. 2.

Chapter 11 Orchestrated Pictures

1. *Patea Mail*, 8 February 1905, p. 2; *Woodville Examiner*, 17 February 1905, p. 2.
2. *Patea Mail*, p. 2; *Wanganui Chronicle*, 15 February 1905, p. 8; *West Coast Times*, 14 April 1905, p. 3; *Ashburton Guardian*, 1 May 1905, p. 2.
3. *New Zealand Free Lance*, 4 March 1905, p. 14
4. *Otago Witness* (*OW*), 27 May 1908, p. 81.
5. Ibid.
6. Stephen Bottomore 'An International Survey of Sound Effects in Early Cinema', *Film History*, 11, 4, 1999, pp. 485–98; Bottomore, 'The Story of Percy Peashaker: Debates about sound effects in the early cinema', in Richard Abel and Rick Altman, eds, *The Sounds of Early Cinema*, pp. 129–42. For other useful accounts, see James Lastra, *Sound Technology and the American Cinema: Perception, representation, modernity*, pp. 104–08; Rick Altman, *Silent Film Sound*, pp. 144–55.
7. *Wanganui Herald*, 29 March 1906, p. 6.
8. *Evening Post* (*EP*), 22 March 1906, p. 4.
9. *Hawera & Normanby Star* (*HNS*), 10 February 1905, p. 2; *West Coast Times*, 15 April 1905, p. 4.
10. David Walker, *Stratford: Shakespearean town under the mountain, a history*, p. 211.
11. *New Zealand Herald* (*NZH*), 26 March 1915, p. 7.
12. Henry J. Hayward, *Here's to Life: The impressions, confessions and garnered thoughts of a free-minded showman*, pp. 81–89; Clive Sowry, 'Hayward, Henry John 1865–1945', Dictionary of New Zealand Biography, www.teara.govt.nz/en/biographies/3h11/1
13. *EP*, 7 July 1905, p. 5.
14. *OW*, 12 April 1905, p. 60; *EP*, 6 July 1905, p. 6; *HNS*, 22 August 1905, p. 1.
15. Maurice Hurst, *Music and the Stage in New Zealand: The record of a century of entertainment*, p. 48.
16. Charles Musser, *The Emergence of Cinema: The American screen to 1907*, pp. 488–89; Richard Abel, '"Pathé Goes to Town": French films create a market for the Nickelodeon', *Cinema Journal*, 25, 1, 1995, pp. 3–26.
17. Brian Robb, *Silent Cinema*, pp. 36–48.

18. *OW*, 12 April 1905, p. 61; *EP*, 7 July 1905, p. 5.
19. Robin Hyde, *Journalese*, p. 176.
20. John A. Lee, *Early Days in New Zealand*, p. 100.
21. Hurst, *Music and the Stage*, p. 48.
22. *HNS*, 22 August 1905, p. 7.
23. P.A. Harrison, 'The Motion Picture Industry in New Zealand 1896–1930: A history of the commercial distribution and exhibition of films', pp. 181–87; Bruce W. Hayward and Selwyn p. Hayward, *Cinemas of Auckland 1896–1979*, pp. 16–19; Nerida Elliott, 'Anzac, Hollywood and Home: Cinemas and film-going in Auckland 1909–1939', pp. 188–91; Squire L. Speedy, *The Picturedrome Fun Merchant: Anecdotes of the life and times of the Picturedrome cinema and dancehall at Milford during the era of L.L. Speedy, 1922–1938*, pp. 17–18; Churchman, *Celluloid Dreams*, pp. 12–13, 27; Emily Thompson, *The Soundscape of Modernity: Architectural acoustics and the culture of listening in America 1900–1933*, pp. 229–84.
24. Henry Shirley, *Just a Bloody Piano Player*, pp. 56–57.
25. Hurst, *Music and the Stage*, p. 68; Altman, *Silent Film Sound*, pp. 289–388.
26. Martin Miller Marks, *Music and the Silent Film: Contexts and case studies 1895–1924*, pp. 93–97; Altman, *Silent Film Sound*, pp. 141, 163, 254, 290, 385.
27. *EP*, 5 July 1913, p. 11; 19 July 1913, p. 3.
28. *EP*, 24 July 1913, p. 8.
29. *EP*, 10 March 1914, p. 3; *Grey River Argus*, 14 August 1914, p. 6.
30. Helen Martin and Sam Edwards, *New Zealand Films 1912–1996*, pp. 21, 23, 38.
31. John Mansfield Thomson, *A Distant Music: The life and times of Alfred Hill 1879–1960*, pp. 57–65.
32. Chris Bourke, *Blue Smoke: The lost dawn of New Zealand popular music 1918–1964*, pp. 9–10.
33. Walker, *Stratford*, p. 211.
34. John Mansfield Thomson, *The Oxford History of New Zealand Music*, p. 161.
35. Speedy, *Picturedrome Fun Merchant*, p. 12.
36. Margery Dixon, 'Early Entertainment in a Mining Town I', *Ohinemuri Regional History Journal*, 19, 1975, p. 20; Jim Sullivan, *Canterbury Voices*, pp. 52–57; Bourke, *Blue Smoke*, pp. 8–9; Kirstine Moffat, *Piano Forte: Stories and soundscapes from colonial New Zealand*, pp. 135–38.
37. Colleen Christie, ed., *Back Then: Oral history interviews from the Birkenhead Public Library Collection Volume 2*, p. 99.
38. Alice O'Callahan, 'I Remember', in Alec Utting, ed., *Life, Laughter and Love in the Early Years: Remembering earlier times in Birkenhead and Birkdale*.
39. Christie, *Back Then, Volume 2*, pp. 34, 115; Christie, *Back Then, Volume 3*, pp. 11–12, 70; Kathy Haddon, *Birkenhead: The way we were*, p. 49.
40. *Wanganui Herald*, 29 December 1905, p. 6.
41. Speedy, *Picturedrome Fun Merchant*, p. 12.
42. Thomson, *Oxford History*, p. 161.
43. For a provocative account of this possibility, see Tim Anderson, 'Reforming "Jackass Music": The problematic aesthetics of early American film music accompaniment', *Cinema Journal*, 37, 1, 1997, pp. 3–22.

44. Lee, *Early Days*, p. 100; Charles Musser, *The Emergence of Cinema: The American screen to 1907*, pp. 361–63; Richard Abel, 'That Most American of Attractions, the Illustrated Song', in Abel and Altman, eds, *Sounds of Early Cinema*, pp. 143–55; Altman, *Silent Film Sound*, pp. 182–93.
45. *West Coast Times*, 6 March 1901, p. 3; *HNS*, 14 March 1905, p. 2; *New Zealand Observer*, 4 April 1908, p. 6.
46. *Grey River Argus*, 9 May 1905, p. 9.
47. *HNS*, 19 December 1908, p. 7.
48. *New Zealand Free Lance*, 30 December 1905, p. 5.
49. *OW*, 24 April 1907, p. 8; 1 May 1907, p. 18. For Lauder's Chronophone recording, see Scott Curtis, 'If It's Not Scottish, It's Crap: Harry Lauder sings for Selig', *Film History*, 11, 4, 1999, pp. 418–25; Altman, *Silent Film Sound*, p. 164.
50. *New Zealand Observer*, 29 June 1907, p. 6; 21 August 1907, p. 6.
51. Abel, 'That Most American of Attractions', p. 144.
52. Reynold Ayers, 'Don'ts For Movie Goers', *New Zealand Theatre & Motion Picture*, 6, May 1921, p. 38.
53. Gregg Bachman, 'Still in the Dark: Silent film audiences', *Film History*, 9, 1, 1997, pp. 23–48; Stephen Bottomore, 'The Panicking Audience?: Early cinema and the train effect', *Historical Journal of Film, Radio and Television*, 19, 2, 1999, pp. 177–216; D.W. Robertson, 'The Noises of Spectators, or the Spectator as Additive to the Spectacle', in Abel and Altman, eds, *Sounds of Early Cinema*, pp. 183–91; Altman, *Silent Film Sound*, pp. 278–85.
54. A.N. Hunt, 'Recreation and Entertainment', in A.N. Hunt, ed., *Foxton 1888–1988: The First 100 Years*, p. 217.
55. Peg Cummins, *A History of Kawhia & Its District*, Kāwhia, 2004, pp. 139–40.
56. Shirley, *Just a Bloody Piano Player*, p. 101.
57. *West Coast Times*, 23 April 1900, p. 4.
58. *NZH*, 2 July 1900, p. 7; *Auckland Star*, 7 July 1900, p. 2; *OW*, 12 September 1900, p. 55; *North Otago Times*, 12 October 1900, p. 2; *Marlborough Express*, 5 December 1900, p. 3; *Grey River Argus*, 21 December 1900, p. 3.
59. Nicholas Reeves, 'Through the Eye of the Camera: Contemporary cinema audiences and their experience of war in the film *Battle of the Somme*', in Hugh Cecil and Peter Liddle, eds, *Facing Armageddon: The First World War experience*, pp. 780–98.
60. Ibid., 26 March 1915, p. 7.

Chapter 12 Talking Pictures

1. Bert Peterson, 'Epi Shalfoon: Loss of popular musician', *Te Ao Hou*, 5, 1953, pp. 19–20; Reo Sheirtcliff, 'Dancing in the Dark: A memoir of Epi Shalfoon', *Music in New Zealand*, 10, 1990, pp. 40–45; Anon., 'Snapping at Our Heels: In the days of dance', *Mana Magazine*, 24, 1998, pp. 60–61; Reo Shalfoon, 'Shalfoon, Gareeb Stephen', Dictionary of New Zealand Biography, www.teara.govt.nz/en/biographies/4s22/1; Robin Hyde, *Disputed Ground: Robin Hyde, journalist*, Wellington, 1991, pp. 240–41; Chris Bourke, *Blue Smoke: The lost dawn of New Zealand popular music 1918–1964*, pp. 61–66; Aleisha Ward, 'New

Zealand's First Jazz Recording', *Audioculture*, www.audioculture.co.nz/scenes/new-zealand-s-first-jazz-recording
2. Diane Pivac, 'The Rise of Fiction: Between the wars', in Diane Pivac, Frank Stark and Lawrence McDonald, eds, *New Zealand Film: An illustrated history*, pp. 68–69.
3. John Archer, 'He Puru Taitama', *New Zealand Folk Song*, www.folksong.org/he_puru/index.html
4. The video can be seen online. See: Ward, 'New Zealand's First Jazz Recording', www.audioculture.co.nz/scenes/new-zealand-s-first-jazz-recording
5. *New Zealand Truth*, 20 August 1931, p. 4.
6. Charles Merrell Bragg, 'Cinema Sings the Blues', *Cinema Journal*, 17, 2, 1978, p. 2.
7. Randy Gue, '"It Seems That Everything Looks Good Nowadays, as Long as it is in the Flesh & Brownskin": The assertion of cultural difference at Atlanta's 81 Theatre, 1934–1937', *Film History*, 8, 2, 1996, pp. 209–18; Katherine Spring, 'Pop Go the Warner Bros., et al.: Marketing film songs during the coming of sound', *Cinema Journal*, 48, 1, 2008, pp. 68–89; Katherine Spring, '"To Sustain Illusion is All That is Necessary": The authenticity of song performance in early American cinema', *Film History*, 23, 3, 2011, pp. 285–99; Rielle Navitski, 'The Tango on Broadway: Carlos Gardel's international stardom and the transition to sound in Argentina', *Cinema Journal*, 51, 1, 2011, pp. 26–49.
8. Donald Crafton, *The Talkies: American cinema's transition to Sound 1926–1931*, pp. 1–18.
9. John Thomson, *The Oxford History of New Zealand Music*, p. 158; James P. Kraft, *Stage to Studio: Musicians and the sound revolution 1890–1950*, pp. 33–58; Bourke, *Blue Smoke*, pp. 43–51.
10. Anon., 'Civic Music: Orchestra, band and organ', *Supplement to the Sun*, 21 December 1929, p. 11.
11. *Auckland Weekly News*, 5 February 1930, p. 63; *Auckland Star* (*AS*), 22 March 1930, p. 5.
12. Caroline Daley, 'Modernity, Consumption and Leisure', in Giselle Byrnes, ed., *The New Oxford History of New Zealand*, pp. 438–39; Charlotte Greenhalgh, 'Bush Cinderellas: Young New Zealanders and romance at the movies 1919–1939', *New Zealand Journal of History*, 44, 1, 2010, pp. 3–4.
13. Gerald Bloomfield, *New Zealand: A handbook of historical statistics*, p. 125.
14. Geoff Lealand, *A Foreign Egg in Our Nest?: American popular culture in New Zealand*, p. 84.
15. Simon Price, *New Zealand's First Talkies: Early film-making in Otago and Southland 1896–1939*, pp. 34–35; Geoffrey Churchman, ed., *Celluloid Dreams: A century of film in New Zealand*, p. 51; David Gerstner and Sarah Greenlees, 'Cinema by Fits and Starts: New Zealand film practices in the twentieth century', *Cineaction*, 51, pp. 39–40; Bruce Babington, *A History of the New Zealand Feature Film*, pp. 50–52; Pivac, 'The Rise of Fiction', pp. 72–76; Diane Pivac, 'Coubray, Edwin', Dictionary of New Zealand Biography, www.teara.govt.nz/en/biographies/5c38/1
16. Price, *New Zealand's First Talkies*, pp. 33–73.; Helen Martin and Sam Edwards, *New Zealand Film 1912–1996*, pp. 45–48.

17. Price, *New Zealand's First Talkies*, pp. 44–52.
18. Ibid., p. 53.
19. Diane Pivac and Robert Sklar, 'Rudall Hayward, New Zealand Film-Maker', *Landfall*, 98, 1971, pp. 147–54; S.R. Edwards, 'Docudrama from the Twenties: Rudall Hayward, Whakatane, and the Te Kooti Trail', *Historical Review*, 41, 2, 1993, pp. 58–63; Churchman, *Celluloid Dreams*, p. 51; Martin and Edwards, *New Zealand Film*, pp. 48–49; Jan Grefstad and Alan Webb, *The Cinemas of Auckland: Volume 1*, pp. 146–50; Babington, *A History of the New Zealand Feature Film*, pp. 55–84; Deborah Shepard, 'Shadow Play: The film-making partnership of Rudall and Ramai Hayward', in Deborah Shepard, ed., *Between the Lives: Partners in art*, pp. 113–35; Sam Edwards and Stuart Murray, 'A Rough Island Life: The Film Life of Rudall Charles Hayward', in Ian Conrich and Stuart Murray, eds, *New Zealand Filmmakers*, pp. 35–53; L.R. Skelton, 'Hayward, Rudall Charles Victor', Dictionary of New Zealand Biography, www.teara.govt.nz/en/biographies/4h22/1; Alistair Fox, 'Rudall Hayward and the Cinema of Maoriland: Genre-mixing and counter-discourses in *Rewi's Last Stand* (1925), *The Te Kooti Trail* (1927) and *Rewi's Last Stand/The Last Stand* (1940)', in Alistair Fox, Barry Keith Grant and Hilary Radner, eds, *New Zealand Cinema: Interpreting the Past*, pp. 45–64; Pivac, 'The Rise of Fiction', pp. 63–72.
20. Deborah Shepard, *Reframing Women: A history of New Zealand film women*, pp. 20–24.
21. Margery Dixon, 'Early Entertainment in a Mining Town II', *Ohinemuri Regional History Journal*, 20, 1976, pp. 29–30; Chris Watson, '*Frances of Fielding* (Lee Hill, 1928) A Community Comedy: New Zealand's populist answer to Hollywood', www.latrobe.edu.au/screeningthepast/firstrelease/fr1199/cwfr8c.htm; Minette Hillyer, 'Dominion Screen Types and Local Beauty Spots: New Zealand's pre-war community films', *New Zealand Journal of Media Studies*, 11, 2, 2008, pp. 37–57; Jeanette Hoorn and Michelle Smith, 'Rudall Hayward's Democratic Cinema and the "Civilising Mission" in the "Land of the Wrong White Crowd"', in Fox, Grant and Radner, *New Zealand Cinema*, pp. 65–81.
22. Hoorn and Smith, 'Rudall Hayward's Democratic Cinema', pp. 68–72.
23. Kristin Thompson and David Bordwell, *Film History: An introduction*, pp. 214–30; Crafton, *The Talkies*, pp. 23–164.
24. Stephen Adams and Orville Butler, *Manufacturing the Future: A History of Western Electric*, Cambridge, 1999, pp. 137–38.
25. Crafton, *The Talkies*, pp. 146–60.
26. Michael Fowler and David Turnbull, *The Reel Story: A history of Napier and Hastings cinemas 1896–1996*, Havelock North, 2008, p. 16.
27. Crafton, *The Talkies*, pp. 51–54.
28. P.A. Harrison, 'The Motion Picture Industry in New Zealand 1896–1930: A history of the commercial distribution and exhibition of films', pp. 213–18.
29. Ibid., pp. 219–20, 223–26.
30. Bloomfield, *Historical Statistics*, p. 25.
31. Geoffrey Churchman, ed., *Celluloid Dreams: A Century of Film in New Zealand*, Wellington, 1997, p. 12; Wayne Brittenden, *Celluloid Circus: The Heyday of the New Zealand Picture Theatre 1925–1970*, Auckland, 2008, p. 11.

32. Harrison, 'Motion Picture Industry in New Zealand 1896–1930', pp. 181–87; Churchman, *Celluloid Dreams*, p. 27; Emily Thompson, *The Soundscape of Modernity: Architectural acoustics and the culture of listening in America 1900–1933*, pp. 245–47.
33. Squire Speedy, *The Picturedrome Fun Merchant: Anecdotes of the life and times of the Picturedrome cinema and dancehall at Milford during the era of L.L. Speedy 1922–1938*, pp. 17–18.
34. *AS*, 24 June 1936, p. 24; 29 June 1936, p. 18.
35. Russell Standish, *Eltham: One hundred years*, p. 74; Karen Christian, ed., *The Eltham Town Hall: Memories of a community treasure*, p. 76.
36. Ibid., p. 10.
37. *Supplement to the Sun*, 21 December 1929, p. 11.
38. Churchman, *Celluloid Dreams*, p. 26; Brittenden, *Celluloid Circus*, pp. 153–57.
39. *AS*, 21 December 1929, p. 12.
40. *The Civic Review Souvenir Number*, December 1929, pp. 24–29.
41. Ibid., p. 22.
42. *AS*, 27 December 1929, p. 5.
43. Richard Barrios, *A Song in the Dark: The birth of the musical film*, pp. 191–93.
44. *AS*, 29 January 1930, p. 28.
45. *AS*, 27 February 1930, p. 28.
46. *AS*, 1 March 1930, p. 21.
47. *Evening Post* (*EP*), 1 September 1932, p. 10.
48. *EP*, 5 September 1932, p. 6.
49. *AS*, 20 March 1930.
50. James Belich, *Paradise Reforged: A history of the New Zealanders from the 1880s to the year 2000*, pp. 251–54.
51. Simon Sigley, 'Film Culture: Its development in New Zealand 1929–1972', p. 50.
52. Simon Sigley, 'How *The Road to Life* (1931) Became the Road to Ruin: The case of the Wellington Film Society in 1933', *New Zealand Journal of History*, 42, 2, 2008, p. 198.
53. Lealand, *Foreign Egg*, pp. 91–92.
54. *AS*, 24 September 1929, p. 8; *EP*, 8 November 1930, p. 23.
55. Nerida Elliott, 'Anzac, Hollywood and Home: Cinemas and film-going in Auckland 1909–1939', pp. 99–101.
56. Dean Scapalo, *The Complete New Zealand Music Charts 1966–2006: Singles, albums, DVDs, compilations*.
57. *The Spike or Victoria College Review*, June 1930, p. 9.
58. John Beaglehole to parents, 28 June 1929, John Cawte Beaglehole Letters, New Zealand Electronic Text Centre, www.nzetc.victoria.ac.nz/tm/scholarly/tei-JCB-081.html
59. *AS*, 2 August 1930, p. 5.
60. Sigley, 'Film Culture', pp. 57–59.
61. Ibid., pp. 63–65, 69–72.
62. *AS*, 11 January 1930, p. 5.
63. Rudolf Arnheim, *Film Essays and Criticism*, p. 30.

64. Crafton, *The Talkies*, pp. 539–46.
65. Roger Openshaw and Roy Shuker, 'Silent Movies and Comics', in Malcolm McKinnon, ed., *The American Connection: Essays from the Stout Centre Conference*, pp. 52–58; Roy Shuker, 'Popular Culture, Moral Panic and Schooling: New Zealand youth at the movies', in John Codd, Richard Harker and Roy Nash, eds, *Political Issues in New Zealand Education*, pp. 235–45; Graham Owen, 'Expectations, Panic and Change: The early years of motion pictures in New Zealand', in Margot Fry, ed., *A Century of Film in New Zealand: Papers from the conference 'Cinema, Film & Society', Wellington, October 1996*, pp. 20–27.
66. Harrison 'Motion Picture Industry', pp. 191–92.
67. *EP*, 9 July 1929, p. 8.
68. *EP*, 11 July 1929, p. 12.
69. Ibid.
70. Harrison 'The Motion Picture Industry', pp. 191–94.
71. *New Zealand Statutes*, 1928, p. 247.
72. *Appendices to the Journals of the House of Representatives*, 1930, H-22, p. 9.
73. *EP*, 28 April 1931, p. 10.
74. Gordon Mirams, *Speaking Candidly: Films and people in New Zealand*, p. 125.
75. *EP*, 17 August 1936, p. 15.
76. Ibid., p. 2.

Conclusion

1. Elizabeth Connors, 'Mechanised Pleasure: Why we get so bored', *Mirror*, 1 September 1931, p. 73.
2. Murray Schafer, *The Soundscape: Our sonic environment and the tuning of the world*, pp. 90–91.
3. Jonathan Sterne, *The Audible Past: Cultural origins of sound reproduction*, pp. 20–21.
4. Shuhei Hosokawa, 'The Walkman Effect', *Popular Music*, 4, 1984, pp. 171–73; Michael Bull, *Sounding Out the City: Personal stereos and the management of everyday life*; Iain Chambers, 'The Aural Walk', in Christoph Cox and Daniel Warner, eds, *Audio Culture: Readings in modern music*, pp. 98–101; Michael Bull, 'Thinking About Sound, Proximity and Distance in Western Experience: The case of Odysseus's walkman', in Veit Erlmann, ed., *Hearing Cultures: Essays on sound, listening and modernity*, pp. 173–91; Steven Levy, *The Perfect Thing: How the iPod shuffles commerce, culture and coolness*; Michael Bull, *Sound Moves: iPod culture and urban experience*.
5. Bull, *Sounding Out the City*, pp. 31–42, 156–61.
6. Marshall Berman, *All That is Solid Melts Into Air: The experience of modernity*, p. 16.
7. Ibid., pp. 15–36; Stephen Toulmin, *Cosmopolis: The hidden agenda of modernity*, pp. 5–44; Anthony Giddens, *Consequences of Modernity*, pp. 1–54; Giddens, *Modernity and Self-Identity: Self and society in the late modern age*, pp. 10–34; Scott Lash and Jonathan Friedman, 'Introduction: Subjectivity and modernity's

other', in Scott Lash and Jonathan Friedman, eds, *Modernity and Identity*, pp. 1–30; Steven Connor, 'The Modern Auditory I', in Roy Porter, ed., *Rewriting the Self: Histories from the Renaissance to the present*, pp. 203–23.
8. Stephen Kern, *The Culture of Time and Space 1880–1918*.
9. J.B. Priestley, *Margin Released: A writer's reminiscences and reflections*, pp. 66–67.
10. Sterne, *The Audible Past*, p. 182.
11. Robert Bresson, *Notes on the Cinematographer*, p. 61.

BIBLIOGRAPHY

•••

1. Primary sources
ARCHIVAL COLLECTIONS
ALEXANDER TURNBULL LIBRARY, WELLINGTON

3ZM Radio Service (Christchurch), Records, MS-Papers-1645.
Albiston, Isobel Shirley, 1921–1995, Diaries and autograph book, MSX-6390.
Anderson, Bruce, 1935-201, Papers, MS-Group-0333.
Anson, Hugo Vernon, 1894–1958, Talks on music, MS-Papers-0327.
Begg, Chas & Co Ltd, Papers, MS-Group-0151.
Broadcasting History Trust, Research notes and papers on Sir James Shelley, MS-Papers-5183.
Buick, Thomas Lindsay, 1865–1938, The romance of the gramophone – Reviews, MS-0351.
Columbia Graphophone (Australia) Ltd, Maori music. Columbia & Parlophone recordings made at Rotorua, N.Z. Columbia Graphophone (Australia) Ltd, Homebush, N.S.W., [ca 1930], Eph-A-PHONO-1930-01.
Covers for 78rpm phonographic records produced by New Zealand recording companies, 1930–1940s, Eph-C-PHONO-COVERS-Hato-1930/40s.
Grainger, Percy Aldridge, 1882–1961, [Catalogue of the] phonograph cylinder collection at the Grainger Museum, Melbourne, MS-Papers-10488.
Huggins, William Percy, fl 1910–1960, Papers relating to the development of radio broadcasting in New Zealand, MS-Papers-6562.
Morgan, Reg, Scrapbooks and address, MS-Papers-4122.
Olive Boyd, interview by Jamie Mackay, 7 February 1992, for the Huntly Coalfields Oral History Project, OHInt-0020/03.
Papers relating to Gladys Watkins, MS-Papers-5666.
Stennett, Gwen, 1893–1982, Papers, MS-Group-1005.
Vink, Patricia Joy, fl 1987, Papers of R.M.J. Haggett concerning broadcasting, MS-Papers-11031.
Wrathall, Dudley H., Scrapbook, MS-Papers-4123.

ARCHIVES NEW ZEALAND, WELLINGTON
New Zealand Broadcasting Service (AAFK)
Correspondence Relating to 'B' Stations, Correspondence Relating to 1ZB, Auckland, 1935–1936, AAFK 890 12c.
Correspondence Relating to 'B' Stations, Station 1ZM, Auckland, 1935–1936, AAFK 890 12d.
Correspondence Relating to Programmes, Music and Plays, Christchurch, 1935–1935, AAFK 890 15b.
Correspondence Relating to Programmes, Programmes, Dunedin, 1934–1934, AAFK 890 17a.

Correspondence Relating to Programmes, Plays and Sketches, 1932–1932, AAFK 890 14a.
Correspondence Relating to Programmes, Wellington Station, Music, 1932–1932, AAFK 890 15a.
Correspondence Relating to Programmes, Wellington Station, Records, 1932–1932, AAFK 890 17b.
Correspondence Relating to Programmes, Wellington Station Programme Matters, 1932–1932, AAFK 890 16b.
Correspondence Relating to Programmes, Workers' Education Association, 1933–1933, AAFK 890 13b.
Correspondence Relating to Special Broadcasts, Madame Zelanda, 1934–1934, AAFK 890 18e.
Other Subjects, Miscellaneous, Auckland, 1935–1935, AAFK 890 23b.
Other Subjects, Plebiscite of Listeners, 1932–1932, AAFK 890 22c.

Broadcasting Corporation of New Zealand, Head Office (AADL)
Aunt Daisy's Session, 1943–1945, AADL 563 3a 2/22 5.
Listener's Survey Board, Plebiscite matters, 1932–1932, AADL 890 W2814 3/d.
Papers of the Broadcasting Coverage Commission, Listeners' Letters, 1931–1938, AADL 890 2e.
Papers of the Broadcasting Coverage Commission, Listener's Survey Board, Plebiscite Matters, 1932–1932, AADL 890 3d.
Radio Coverage, Dora Lindsay, 1931–1932, AADL 890 2c.
Shirley, Henry, 1936–1967, AADL 793 W2814 28 Alpha 'S'.
Talks, A.E.S. [Adult Educational Service?], 1937 Dunedin Talks, 1936–1937, AADL 955 4d.
Trade Claims Correspondence with Public, Criticisms, etc. Suggestions, 1936–1937, AADL 564 95a 2/1/1 1.

GEORGE GREY SPECIAL COLLECTIONS, AUCKLAND LIBRARIES

Grey, George, Grey New Zealand letters, GLNZ S11.1; GLNZ O11.6; GLNZ U1; GLNZ W33.
Grefstad, Jan and Allan Webb, *Auckland Cinemas*, 2 volumes, unpaged, 2002.
Grefstad, Jan, *Notes on Auckland Cinemas*, NZMS 1114.
McWilliams, Dave, *Auckland's Film History 1896–1941*, Auckland, 1941.
Zalinsky, Rona von, *Civic Theatre Scrapbook*, NZMS 1386.

RADIO NEW ZEALAND NATIONAL SOUND ARCHIVE, CHRISTCHURCH

Aunt Gwen of 2YA with Uncle Bruce, Id Number 33551.
D Series: Portrait From Life: Clive Drummond, D1032 1–2.

2. Secondary sources

A. BOOKS, ARTICLES AND THESES

Abel, Richard, '"Pathé Goes to Town": French films create a market for the nickelodeon', *Cinema Journal*, 25, 1, 1995, pp. 3–26.

Abel, Richard, 'That Most American of Attractions, the Illustrated Song', in Richard Abel and Robert Altman, eds, *The Sounds of Early Cinema*, Bloomington, 2001, pp. 143–55.

Abel, Richard, 'The Cinema of Attractions in France 1896–1904', in Lee Grieveson and Peter Krämer, eds, *The Silent Cinema Reader*, Oxford, 2006, pp. 63–75.

Abel, Richard and Rick Altman, eds, *The Sounds of Early Cinema*, Bloomington, 2001.

Ackerman, Diane, *A Natural History of the Senses*, New York, 1981.

Adams, Stephen and Orville Butler, *Manufacturing the Future: A history of Western Electric*, Cambridge, 1999.

Adorno, Theodor, *Essays on Music*, edited by Richard Leppert, Berkeley, CA, 2002.

Adorno, Theodor, 'Analytical Study of the NBC Music Appreciation Hour', *The Musical Quarterly*, 28, 2, 1994, pp. 325–77.

Adorno, Theodor, 'The Curves of the Needle', in Richard Leppert, ed., *Essays on Music*, pp. 271–76.

Adorno, Theodor, 'The Form of the Phonograph Record', in Richard Leppert, ed., *Essays on Music*, pp. 277–82.

Adorno, Theodor, 'On the Fetish-Character in Music and the Regression of Listening', in Richard Leppert, ed., *Essays on Music*, pp. 288–317.

Adorno, Theodor, 'On Popular Music', in Richard Leppert, ed., *Essays on Music*, pp. 437–69.

Adorno, Theodor, 'Opera and the Long Playing Record', in Richard Leppert, ed., *Essays on Music*, pp. 283–87.

Adorno, Theodor, 'The Radio Symphony', in Richard Leppert, ed., *Essays on Music*, pp. 251–70.

Adorno, Theodor and Max Horkheimer, *Dialectic of Enlightenment*, London, 1997.

Alexandratos, Michael, 'The Rotorua Maori Choir: A history, discography and analysis, *The Discographer Magazine*, 3, 6, 2016, pp. 3–24.

Altman, Rick, 'Introduction', *Yale French Studies Cinema/Sound*, 60, 1980, pp. 3–15.

Altman, Rick, 'The Silence of the Silents', *The Musical Quarterly*, 80, 4, 1996, pp. 648–718.

Altman, Rick, ed., *Sound Theory, Sound Practice*, New York, 1992.

Altman, Rick, *Silent Film Sound*, New York, 2004.

Anderson, Atholl, Judith Binney and Aroha Harris, *Tangata Whenua: An illustrated history*, Wellington, 2014.

Anderson, Barbara, *Getting There: An autobiography*, Wellington, 2009.

Anderson, Benedict, *Imagined Communities: Reflections on the origins and spread of nationalism*, New York, 1991.

Anderson, Bruce, *Story of the New Zealand Record Industry*, Wellington, 1984.

Anderson, Tim, 'Reforming "Jackass Music": The problematic aesthetics of early American film music accompaniment', *Cinema Journal*, 37, 1, 1997, pp. 3–22.

Annabell, N. (ed), *Official History of the New Zealand Engineers during the Great War, 1914–1919: A record of the work carried out by the field companies, field troops, Signal Troop and Wireless Troop, during the operations in Samoa (1914–15); Egypt, Gallipoli, Sinai and Palestine (1914–1918); France, Belgium and Germany (1916–1919); and Mesopotamia (1916–1918)*, Wanganui, 1927.
Anon, 'The Singing Tahiwis', *Otaki Historical Society Journal*, 21, 1998, pp. 45–49.
Armstrong, Allen, 'Still Popular after Thirty Years: How Rotorua club made first recordings of Maori music', *Te Ao Hou*, 36, 1961, pp. 63–64.
Armstrong, Alan, 'The Great Songs of Ana Hato and Deane Waretini', *Te Ao Hou*, 44, 1963, pp. 57–58.
Armstrong, Tim, 'Player Piano: Poetry and sonic modernity', *Modernism/Modernity*, 14, 1, 2007, pp. 1–19.
Arnold, Rollo, 'The Country Child in Later Victorian New Zealand', in H. Debenham and W. Slinn, eds, *Australasian Victorian Studies Association: Conference papers*, Christchurch, 1981, pp. 1–15.
Ashby, Arved, *Absolute Music, Mechanical Reproduction*, Berkeley, CA, 2010.
Ashbrook, William, 'Maurice Renaud: The complete gramophone recordings 1901–1908', *The Opera Quarterly*, 15, 2, 1999, pp. 324–27.
'Atom', 'Wireless', *The Kia Ora Coo-Ee: The magazine for the Anzacs in the Middle East, 1918*, Sydney, Facsimile, 1981, Second Series, p. 2.
Attali, Jacques, *Noise: The political economy of music*, Minneapolis, 1985.
Audio Devices, Inc., *How to Make Good Recordings: The complete handbook for the everyday recordist*, 2nd edn, New York, 1948.
Augoyard, Jean-François and Henry Torgue, eds, *Sonic Experience: A guide to everyday sounds*, Montreal, 2005.
Auslander, Philip, 'Looking at Records', *The Drama Review*, 45, 1, 2001, pp. 77–83.
Babington, Bruce, *A History of the New Zealand Feature Film*, Manchester, 2007.
Bachman, Gregg, 'Still in the Dark: Silent film audiences', *Film History*, 9, 1, 1997, pp. 23–48.
Bailey, Peter, *Leisure and Class in Victorian England: Rational recreation and the contest for control 1830–1885*, London, 1978.
Bailey, Peter, ed., *Music Hall: The business of pleasure*, Milton Keynes, 1986.
Bailey, Peter, 'Breaking the Sound Barrier: A historian listens to noise', *Body & Society*, 2, 2, 1996, pp. 49–66.
Bailey, Peter, *Popular Culture and Performance in the Victorian City*, Cambridge, 1998.
Bailey, Peter, 'The Politics and Poetics of Modern British Leisure: A late twentieth-century review', *Rethinking History*, 3, 2, 1999, pp. 131–75.
Ball, Desmond, Cliff Lord and Meredith Thatcher, *Invaluable Service: The secret history of New Zealand's signals intelligence during two world wars*, Waimauku, 2011, pp. 3–6.
Ballantyne, Tony and Brian Montgomery, 'Angles of Vision', in Tony Ballantyne and Brian Montgomery, eds, *Disputed Histories: Imagining New Zealand's pasts*, Dunedin, 2006, pp. 1–24.
Ballantyne, Tony and Brian Montgomery, eds, *Disputed Histories: Imagining New Zealand's pasts*, Dunedin, 2005.

Barfe, Louis, *Where Have All the Good Times Gone?: The rise and fall of the record industry*, London, 2004.

Barnett, Kyle S., 'Furniture Music: The phonograph as furniture 1900–1930', *Journal of Popular Music Studies*, 18, 3, 2006, pp. 301–24.

Barnett, Kyle S., 'The Recording Industry's Role in Media History', in Janet Staiger and Sabine Hake, eds, *Convergence Media History*, New York, 2009, pp. 81–91.

Baron, David S., 'Noise and Degeneration: Theodor Lessing's crusade for quiet', *Journal of Contemporary History*, 17, 1, 1982, pp. 165–78.

Barrios, Richard, *A Song in the Dark: The birth of the musical film*, New York, 1995.

Barrowman, Rachel, *A Popular Vision: The arts and the left in New Zealand 1930–1950*, Wellington, 1991.

Barrowman, Rachel, *Victoria University of Wellington 1899–1999: A history*, Wellington, 1999.

Barthes, Roland, 'The Grain of the Voice', in *Image–Music–Text*, London, 1977, pp. 179–89.

Barthes, Roland, *Image–Music–Text*, London, 1977.

Barthes, Roland, *Mythologies*, St Albans, 1972.

Baume, Eric, *I Lived These Years*, London, 1941.

Baxendale, John, 'Into Another Kind of Life in which Anything Might Happen: Popular music and late modernity 1910–1930', *Popular Music*, 14, 2, 1995, pp. 137–54.

Baxter, Archibald, *We Will Not Cease*, Christchurch, 1965.

Beck, Jay and Tony Grajeda, eds, *Lowering the Boom: Critical studies in film sound*, Urbana, IL, 2008.

Beck, Jay and Tony Grajeda, 'Introduction: The future of film Sound Studies', in Jay Beck and Tony Grajeda, eds, *Lowering the Boom: Critical studies in film sound*, Urbana, IL, 2008, pp. 1–20.

Begg, Charles and Sons, *Begg's Jubilee Souvenir 1861–1911*, Dunedin, 1911.

Belich, James, *Paradise Reforged: A history of the New Zealanders from the 1880s to the year 2000*, Auckland, 2001.

Benjamin, Walter, 'The Work of Art in the Age of Mechanical Reproduction', in H. Arendt, ed., *Illuminations*, London, 1973, pp. 211–44.

Benjamin, Walter, *The Work of Art in the Age of its Mechanical Reproducibility and Other Writings on the Media*, Michael W. Jennings, Brigid Doherty and Thomas Y. Levin, eds, Cambridge, MA, 2008.

Bennett, Arnold, Barry Shank and Jason Toynbee, eds, *The Popular Music Studies Reader*, London, 2006.

Bergh, Arild and Tia DeNora, 'From Wind-up to iPod: Techno-cultures of listening', in Nicholas Cook, Eric Clarke, Daniel Leech-Wilkinson and John Rink, eds, *The Cambridge Companion to Recorded Music*, Cambridge, 2009, pp. 102–15.

Bergonzi, Benet, *Old Gramophones and Other Talking Machines*, Princes Risborough, 1995.

Berman, Marshall, *All That is Solid Melts Into Air: The experience of modernity*, 2nd edn, New York, 1988.

Bijsterveld, Karin, 'The City of Din: Decibels, noise and neighbors in the Netherlands 1910–1980', *Osiris*, 18, 1, 2003, pp. 173–93.

Bijsterveld, Karin, 'The Diabolical Symphony of the Mechanical Age: Technology and symbolism of sound in European and North American noise abatement campaigns 1900–40', *Social Studies of Science*, 31, 1, 2001, pp. 37–70.

Bijsterveld, Karin, *Mechanical Sound: Technology, culture, and public problems of noise in the twentieth century*, Cambridge, MA, 2008.

Bijsterveld, Karin and José van Dijck, 'Introduction', in Karin Bijsterveld and José van Dijck, eds, *Sound Souvenirs: Audio technologies, memory and cultural practices*, Amsterdam, 2009, pp. 11–21.

Billinge, Mark, 'A Time and Place for Everything: An essay on recreation, re-creation and the Victorians', *Journal of Historical Geography*, 22, 4, 1996, pp. 443–59.

Biocca, Frank, 'Media and Perceptual Shifts: Early radio and the clash of musical cultures', *Journal of Popular Culture*, 24, 2, 1990, pp. 1–15.

Biocca, Frank, 'The Pursuit of Sound: Radio, reception and utopia in the early twentieth century', *Media, Culture & Society*, 10, 1, 1988, pp. 61–79.

Birdsall, Carolyn, *Nazi Soundscapes: Sound, technology and urban space in Germany, 1933–1945*, Amsterdam, 2012.

Birdsall, Carolyn and Anthony Enns, eds, *Sonic Meditations: Body sound technology*, Cambridge, 2008.

Birkenmaier, Anke, 'From Surrealism to Popular Art: Paul Deharme's radio theory', *Modernism/Modernity*, 16, 2, 2009, pp. 357–74.

Blake, Eric C., *Wars, Dictators and the Gramophone 1898–1945*, York, 2004.

Blanks, Harvey, *The Golden Road: A record collector's guide to music appreciation*, Sydney, 1968.

Blanning, Tim, *The Triumph of Music: Composers, musicians and their audiences 1700 to the present*, London, 2008.

Bloomfield, Gerald, *New Zealand: A handbook of historical statistics*, Boston, MA, 1984.

Bordwell, David, Janet Staiger and Kristin Thompson, *The Classical Hollywood Cinema: Film style & mode of production to 1960*, London, 1985.

Botstein, Leon, 'Toward a History of Listening', *The Musical Quarterly*, 82, 3/4, 1998, pp. 427–31.

Bottomore, Stephen, 'The Panicking Audience?: Early cinema and the train effect', *Historical Journal of Film, Radio and Television*, 19, 2, 1999, pp. 177–216.

Bottomore, Stephen, 'An International Survey of Sound Effects in Early Cinema', *Film History*, 11, 4, 1999, pp. 485–98.

Bottomore, Stephen, 'The Story of Percy Peashaker: Debates about sound effects in the early cinema', in Richard Abel and Rick Altman, eds, *The Sounds of Early Cinema*, Bloomington, 2001, pp. 129–42.

Bourke, Chris, *Blue Smoke: The lost dawn of New Zealand popular music 1918–1964*, Auckland, 2010.

Boyle, Andrew, *Only the Wind Will Listen: Reith of the BBC*, London, 1972.

Braatvedt, Susan, 'A History of Music Education in New Zealand State Primary and Intermediate Schools 1878–1989', PhD thesis, University of Canterbury, 2002.

Braatvedt, Susan, 'Douglas Tayler: Man of action and initiative', *Sound Ideas*, 6, 1, 2003, pp. 6–18.

Braatvedt, Susan, 'A Few Musical Memories', *Sound Ideas*, 6, 1, 2003, pp. 54–56.
Bradwell, Cyril R., *Fight the Good Fight: The story of the Salvation Army in New Zealand 1883–1983*, Wellington, 1982.
Brady, Erika, *A Spiral Way: How the phonograph changed ethnography*, Jackson, 1999.
Bragg, Charles Merrell, 'Cinema Sings the Blues', *Cinema Journal*, 17, 2, 1978, pp. 1–12.
Brailsford, Ian, 'Enlightened Buying: From consumer service to the Consumers' Institute', *New Zealand Journal of History*, 41, 2, 2007, pp. 123–42.
Brailsford, Ian, 'If There's Not One Near You Now, There Soon Will Be: American fast-food chains come to New Zealand', *New Zealand Journal of History*, 39, 1, 2005, pp. 57–74.
Brickell, Chris, 'The Politics of Post-War Consumer Culture', *New Zealand Journal of History*, 40, 2, 2006, pp. 133–55.
Briggs, Asa, *The Birth of Broadcasting: The history of broadcasting in the United Kingdom Volume I*, London, 1961.
Briggs, Asa and Peter Burke, *A Social History of Media: From Gutenberg to the internet*, Cambridge, 2002.
Brittenden, Wayne, *Celluloid Circus: The heyday of the New Zealand picture theatre 1925–1970*, Auckland, 2008.
Brock-Nannestad, George, 'The Development of Recording Technologies', in Nicholas Cook, Eric Clarke, Daniel Leech-Wilkinson and John Rink, eds, *The Cambridge Companion to Recorded Music*, Cambridge, 2009, pp. 149–76.
Brookes, Barbara, Annabel Cooper and Robin Law, eds, *Sites of Gender: Women, men and modernity in southern Dunedin 1890–1939*, Auckland, 2003.
Brookes, Barbara, Erik Olssen and Emma Beer, 'Spare Time?: Leisure, gender and modernity', in Barbara Brookes, Annabel Cooper and Robin Law, eds, *Sites of Gender: Women, men and modernity in Southern Dunedin 1890–1939*, Auckland, 2003, pp. 151–89.
Brooks Tim, 'George W. Johnson: The first African-American recording star', *ARSC Journal*, 35, 1, 2004, pp. 37–66.
Brooks Tim, *Lost Sounds: Blacks and the birth of the recording industry 1890–1919*, Urbana, IL, 2004.
Brown, C. Mackenzie, 'Purāna as Scripture: From sound to image of the holy word in the Hindu tradition', *History of Religions*, 26, 1, 1986, pp. 68–86.
Brown, Lee B., 'Phonography, Repetition and Spontaneity', *Philosophy and Literature*, 24, 1, 2000, pp. 111–25.
Brown, Richard and Barry Anthony, *A Victorian Film Enterprise: The history of the British Mutoscope and Biograph Company 1897–1915*, Trowbridge, 1999.
Buick, T. Lindsay, *The Romance of the Gramophone*, Wellington, 1927.
Bull, Michael, *Sound Moves: iPod culture and urban experience*, London, 2007.
Bull, Michael, *Sounding Out the City: Personal stereos and the management of everyday life*, Oxford, 2000.
Bull, Michael, 'Thinking About Sound, Proximity and Distance in Western Experience: The case of Odysseus's walkman', in Veit Erlmann, ed., *Hearing Cultures: Essays on sound, listening and modernity*, Oxford, 2004, pp. 173–91.
Bull, Michael and Les Black, eds, *The Auditory Culture Reader*, London, 2003.

Burnett, Charles, Michael Fend and Penelope Gouk, eds, *The Second Sense: Studies in hearing and musical judgment from antiquity to the seventeenth century*, London, 1991.
Burns, Christopher, 'Parading Kiwis: New Zealand soldier concert parties, 1916–1954', MA thesis, University of Auckland, 2012.
Burstyn, Shai, 'In Quest of the Period Ear', *Early Music*, 25, 4, 1997, pp. 693–703.
Burstyn, Shai, 'Music as Heard', *Early Music*, 26, 3, 1998, pp. 515, 517–18.
Burstyn, Shai, 'Pre-1600 Music Listening: A methodological guide', *The Musical Quarterly*, 82, 3/4, 1998, pp. 455–65.
Butsch, Richard, 'Crystal Sets and Scarf-Pin Radios: Gender, technology and the construction of American radio listening in the 1920s', *Media, Culture & Society*, 20, 4, 1998, pp. 557–72.
Bynum, W.F. and Roy Porter, eds, *Medicine and the Five Senses*, Cambridge, 1993.
Byrnes, Giselle, 'Introduction: Reframing New Zealand History', in Giselle Byrnes, ed., *The New Oxford History of New Zealand*, Melbourne, 2009, pp. 1–18.
Byrnes, Giselle, ed., *The New Oxford History of New Zealand*, Melbourne, 2009.
Cameron, Ewan K., 'Obituary: Lucy May Cranwell 1907–2000', *New Zealand Journal of Botany*, 38, 2000, pp. 527–35.
Camlot, Jason, 'Early Talking Books: Recitation anthologies 1880–1920', *Book History*, 6, 2003, pp. 147–73.
Carter, Ian, *Gadfly: The Life and Times of James Shelley*, Auckland, 1993.
Carter, Paul, *The Sound In-Between: Voice, space, performance*, Kensington, NSW, 1992.
Cecil, Hugh and Peter Liddle, eds, *Facing Armageddon: The First World War experience*, London, 1996.
Chambers, Iain, 'The Aural Walk', in Christoph Cox and Daniel Warner, eds, *Audio Culture: Readings in modern music*, New York, 2004, pp. 98–101.
Chanan, Michael, *From Handel to Hendrix: The composer in the public sphere*, London, 1999.
Chanan, Michael, *Musica Practica: The social practice of western music from Gregorian chant to postmodernism*, London, 1994.
Chanan, Michael, *Repeated Takes: A short history of recording and its effects on music*, London, 1995.
Charosh, Paul, 'Popular and Classical in the Mid-Nineteenth Century', *American Music*, 10, 2, 1992, pp. 117–35.
Chew, V.K., *Talking Machines*, London, 1981.
Chion, Michel, *Audio-Visual: Sound on screen*, New York, 1990.
Chow, Rey and James A. Steintrager, 'In Pursuit of the Object of Sound', *Differences: A journal of feminist studies, the sense of sound*, 22, 2–3, 2011, pp. 1–9.
Christian, Karen, ed., *The Eltham Hall: Memories of a community treasure*, Stratford, NZ, 2003.
Christie, Colleen, ed., *Back Then: Oral history interviews from the Birkenhead Public Library collection, Volume 2*, Birkenhead, 1988, *Volume 3*, Birkenhead, 1989.
Christie, Ian, 'Early Phonograph Culture and Moving Pictures', in Richard Abel and Rick Altman, eds, *The Sounds of Early Cinema*, Bloomington, 2001, pp. 3–12.

Churchman, Geoffrey B., ed., *Celluloid Dreams: A century of film in New Zealand*, Wellington, 1997.
Clark, Garry E., 'The Noise of Our Time: Our mistreatment of music in the twentieth century', *Arts Education Policy Review*, 94, 3, 1993, pp. 2–6.
Clarke, Alison, *Holiday Seasons: Christmas, New Year and Easter in nineteenth century New Zealand*, Auckland, 2007.
Clarke, Eric F., 'The Impact of Recording on Listening', *Twentieth-Century Music*, 4, 1, 2007, pp. 47–70.
Clarke, Eric F., *Ways of Listening: An ecological approach to the perception of musical meaning*, Oxford, 2005.
Cleveland, Les, 'What They Liked: Movies and modernity Downunder', *Journal of Popular Culture*, 36, 4, 2003, pp. 756–79.
Clover, Ken, ed., *Hauraki Memories: Some treasured memories from the wider Hauraki area*, Ngatea, 2008.
Clover, Ken, ed., *The People of the Plains: Some early memories of those who have lived in or been associated with the Hauraki Plains*, 3 volumes, Ngatea, 2004.
Cockayne, Emily, *Hubbub: Filth, noise & stench in England 1600–1770*, London, 2007.
Codd, John, Richard Harker and Roy Nash, eds, *Political Issues in New Zealand Education*, 2nd edn, Palmerston North, 1990.
Colles, H.C., *Walford Davies: A biography*, London, 1942.
Collins, Ken G., *Broadcasting Grave and Gay*, Christchurch, 1967.
Connor, Steven, *Dumbstruck: A cultural history of ventriloquism*, Oxford, 2000.
Connor, Steven, 'The Modern Auditory I', in Roy Porter, ed., *Rewriting the Self: Histories from the Renaissance to the present*, London, 1997, pp. 203–23.
Conrich, Ian and Stuart Murray, eds, *New Zealand Filmmakers*, Detroit, 2007.
Cook, Nicholas, Eric Clarke, Daniel Leech-Wilkinson and John Rink, eds, *The Cambridge Companion to Recorded Music*, Cambridge, 2009.
Cooke, Abigail, 'Humorous Reflections on Laughing Records', *ARSC Journal*, 32, 2, 2001, pp. 232–42.
Cooper, Dave, *The Perfect Portable Phonograph*, London, 2003.
Copley, Antony, 'Music in the Himalayas: Alexander Scriabin and the spiritual', *Studies in History*, 26, 2, 2010, pp. 211–26.
Corbin, Alain, *Time, Desire, and Horror: Towards a history of the senses*, Cambridge, MA, 1995.
Corbin, Alain, *Village Bells: Sound and meaning in the 19th-century French countryside*, New York, 1998.
Cousins, Mark, *The Story of Film*, New York, 2004.
Cowan, James, *Official Record of the New Zealand International Exhibition of Arts and Industries, Held at Christchurch 1906–7: A descriptive and historical account*, Wellington, 1910.
Cox, Christoph and Daniel Warner, eds, *Audio Culture: Readings in modern music*, New York, 2004.
Cox, Gordon, '"Changing the Face of School Music": Walford Davies, the gramophone and the radio', *British Journal of Music Education*, 14, 1, 1997, pp. 45–55.

Cox, Gordon, *Living Music in Schools 1923–1999: Studies in the history of music education in England*, Aldershot, 2002.
Crafton, Donald, *The Talkies: American Cinema's Transition to Sound 1926–1931*, Berkeley, CA, 1999.
Crangle, Richard, '"Next Slide Please": The lantern lecture in Britain 1890–1910', in Richard Abel and Rick Altman, eds, *The Sounds of Early Cinema*, Bloomington, 2001, pp. 39–47.
Crary, Jonathan, *Suspensions of Perception: Attention, spectacle, and modern culture*, Cambridge, MA, 1999.
Crary, Jonathan, *Techniques of the Observer: On vision and modernity in the nineteenth century*, Cambridge, MA, 1992.
Crawford, Kate, 'Four Ways of Listening with an iPhone: From sound and network listening to biometric data and geolocative tracking', in Larissa Hjorth, Jean Burgess and Ingrid Richardson, eds, *Studying Mobile Media: Cultural technologies, mobile communication and the iPhone*, New York, 2012, pp. 213–28.
Crowhurst, Andrew, 'The Portly Grabbers of 75 per cent: Capital investment in the British entertainment industry 1885–1914', *Leisure Studies*, 20, 2, 2001, pp. 107–23.
Crowley, T.E., *Discovering Mechanical Music*, Princes Risborough, 1977.
Cuddy-Keane, Melba, 'Virginia Woolf, Sound Technologies and the New Aurality', in Pamela L. Gaughie, ed., *Virginia Woolf in the Age of Mechanical Production*, New York, 2000, pp. 69–96.
Cummins, Peg, *A History of Kawhia & Its District*, Kawhia, 2004.
Curtis, Scott, 'If It's Not Scottish, It's Crap: Harry Lauder sings for Selig', *Film History*, 11, 4, 1999, pp. 418–25.
Daley, Caroline, *Girls & Women, Men & Boys: Gender in Taradale 1886–1930*, Auckland, 1999.
Daley, Caroline, *Leisure & Pleasure: Reshaping and revealing the New Zealand body 1900–1960*, Auckland, 2003.
Daley, Caroline, 'Modernity, Consumption and Leisure', in Giselle Byrnes, ed., *The New Oxford History of New Zealand*, Melbourne, 2009, pp. 423–47.
Damousi, Joy, '"The Filthy American Twang": Elocution, the advent of American "talkies", and Australian cultural identity', *The American Historical Review*, 112, 2, 2007, pp. 394–416.
Damousi, Joy and Desley Deacon, eds, *Talking and Listening in the Age of Modernity: Essays on the history of sound*, Canberra, 2007.
Damousi, Joy and Paula Hamilton, eds, *A Cultural History of Sound, Memory, and the Senses*, New York, 2017.
Danius, Sara, 'Novel Visions and the Crisis of Culture: Visual technology, modernism and death in *The Magic Mountain*', *Boundary*, 27, 2, 2000, pp. 177–211.
Davis, C.S., 'Service Without Sales: The dealer's point of view again', *Gramophone*, 6, 72, 1929, pp. 526–82.
Day, Patrick, *The Radio Years: A history of broadcasting in New Zealand Volume One*, Auckland, 1994.

Day, Patrick, 'American Popular Culture and New Zealand Broadcasting: The reception of early radio serials', *Journal of Popular Culture*, 30, 1, 1996, pp. 203–14.

Day, Patrick, *Voice & Vision: A history of broadcasting in New Zealand Volume Two*, Auckland, 2000.

Day, Patrick, 'Broadcasting the Centennial', in William Renwick, ed., Creating a National Spirit: Celebrating New Zealand's Centennial, Wellington, 2004, pp. 77–86.

Day, Timothy, *A Century of Recorded Music: Listening to musical history*, New Haven, 2000.

Dennis, Jonathan, 'Ana Hato: The melody is ended but the memory lingers on', *Music in New Zealand*, 31, 1995/96, pp. 46–47.

Dennis, Jonathan and Jan Bieringa, eds, *Film in Aotearoa New Zealand*, Wellington, 1992.

DeNora, Tia, *Music in Everyday Life*, Cambridge, 2000.

Department of Education, *Syllabus of Instruction for Public Schools*, Wellington, 1928.

DeVeaux, Scott, 'Constructing the Jazz Tradition: Jazz historiography', *Black American Literature Forum*, 25, 3, 1991, pp. 525–60.

Dickson, W.K.L. and Antonia Dickson, *The Life and Inventions of Thomas Alva Edison*, London, 1894.

Dickson, W.K.L. and Antonia Dickson, *History of the Kinetograph, Kinetoscope and Kineto-Phonograph*, 1895, New York, facsimile, 2000.

Dinerstein, Joel, *Swinging the Machine: Modernity, technology, and African American culture between the two world wars*, Amherst, 2003.

Dix, John, *Stranded in Paradise: New Zealand rock'n'roll 1955 to the modern era*, 2nd edn, Auckland, 2005.

Dixon, Margery, 'Early Entertainment in a Mining Town I', *Ohinemuri Regional History Journal*, 19, 1975, pp. 17–21.

Dixon, Margery, 'Early Entertainment in a Mining Town II', *Ohinemuri Regional History Journal*, 20, 1976, pp. 27–32.

Dorsey, Colin, *Dance Bands of the 20th Century in North Otago*, Tauranga, 2001.

Dougherty, Ian, *Ham Shacks, Brass Pounders & Rag Chewers: A history of amateur radio in New Zealand*, Wellington, 1997.

Douglas, Susan J., *Inventing American Broadcasting 1899–1922*, Baltimore, 1987.

Douglas, Susan J., *Listening In: Radio and the American imagination*, Minneapolis, 2004.

Dow, Derek, 'The Long Locum: Health propaganda in New Zealand', *New Zealand Medical Journal*, 116, 1170, 2003, pp. 1–6.

Dowd, Timothy J., 'Culture and Commodification: Technology and structural power in the early US recording industry', *International Journal of Sociology and Social Policy*, 22, 1, 2002, pp. 106–40.

Downes, Peter, *Top of the Bill: Entertainers through the years*, Wellington, 1979.

Downes, Peter, *The Pollards: A family and its child and opera companies in New Zealand and Australia 1880–1910*, Wellington, 2002.

Downes, Peter and Peter Harcourt, *Voices in the Air: Radio broadcasting in New Zealand: A documentary*, Wellington, 1976.

Doyle, Peter, *Echo and Reverb: Fabricating space in popular music recording 1900–1960,* Middletown, CT, 2005.
Dreyfus, L., 'Early Music Defended against its Devotees: A theory of historical performance in the twentieth century', *The Musical Quarterly,* 69, 1983, pp. 297–322.
Drew, H.T.B., *The War Effort of New Zealand,* Auckland, 1923.
Drobnick, Jim, ed., *Aural Cultures,* Toronto, 2004.
Dym, Jeffrey A., 'Benshi and the Introduction of Motion Picture in Japan', *Monumenta Nipponica,* 55, 4, 2000, pp. 509–36.
East, Gavin, 'Edison Diamond Discs: New Zealand perspective', *The Phonographic Record,* 36, 2, 2001, pp. 37–39.
Eckersley, P.P., *All About Your Wireless Set,* London, 1925.
Edge, Ruth, 'Mr Bohanna Goes Down Under: Parts 1–3', *Australian Record and Music Review,* 7, 1990, pp. 12–15; 10, 1991, pp. 12–16; 17, 1993, pp. 3–5.
Edison, Thomas, 'The Phonograph and its Future', *North American Review,* 126, 262, 1878, pp. 527–36.
Edison, Thomas, 'The Perfected Phonograph', *North American Review,* 146, 379, 1888, pp. 641–50.
Edwards, Les, *Scrim: Radio rebel in retrospect,* Auckland, 1971.
Edwards, S.R., 'Docudrama from the Twenties: Rudall Hayward, Whakatane, and the Te Kooti Trail', *Historical Review,* 41, 2, 1993, pp. 58–63.
Edwards, Sam and Stuart Murray, 'A Rough Island Story: The film life of Rudall Charles Hayward', in Ian Conrich and Stuart Murray, eds, *New Zealand Filmmakers,* Detroit, 2007, pp. 35–53.
Eggleton, David, *Ready to Fly: The story of New Zealand rock music,* Nelson, 2003.
Eglash, Ron, Jennifer L. Croissant, Giovanna Di Chiro, Rayvon Fouché et al., eds, *Appropriating Technology: Vernacular Science and Social Power,* Minneapolis, 2004.
Eisenberg, Evan, *The Recording Angel: The experience of music from Aristotle to Zappa,* New York, 1987.
Elliott, Nerida, 'Anzac, Hollywood and Home: Cinemas and film-going in Auckland 1909–1939', MA thesis, University of Auckland, 1989.
Ellwood, R.S., *Islands of the Dawn: The story of alternative spirituality in New Zealand,* Honolulu, 1993.
Enns, Anthony, 'Psychic Radio: Sound technologies, ether bodies and spiritual vibrations', *Senses & Society,* 3, 2, 2008, pp. 137–52.
Enns, Anthony, 'Voices of the Dead: Transmission translation transgression', *Culture, Theory & Critique,* 46, 1, 2005, pp. 11–27.
Erlmann, Veit, 'But What of the Ethnographic Ear? Anthropology, sound and the senses', in Veit Erlmann, ed., *Hearing Cultures: Essays on sound, listening and modernity,* Oxford, 2002, pp. 1–20.
Erlmann, Veit, *Reason and Resonance: A history of modern aurality,* New York, 2010.
Erlmann, Veit, ed., *Hearing Cultures: Essays on sound, listening and modernity,* Oxford, 2004.
Ettema, James D. and Charles D. Whitney, eds, *Audiencemaking: How the media create the audience,* Thousand Oaks, CA, 1994.

Eyman, Scott, *The Speed of Sound: Hollywood and the talkies revolution 1926–1930*, New York, 1997.

Fabrizio, Timothy C., 'Ordinary People: The talking machine in real life', *ARSC Journal*, 30, 1, 1999, pp. 20–25.

Fabrizio, Timothy C. and George F. Paul, *Antique Phonograph Advertising: An illustrated history*, Atglen, PA, 2002.

Fairburn, Miles, 'Is There a Good Case for New Zealand Exceptionalism?', in Tony Ballantyne and Brian Montgomery, eds, *Disputed Histories: Imagining New Zealand's pasts*, Dunedin, 2005, pp. 143–67.

Farrell, Gerry, 'The Early Days of the Gramophone Industry in India: Historical, social and musical perspectives', *British Journal of Ethnomusicology*, 3, 1993, pp. 31–52.

Faulkner, Anne Shaw, *What We Hear In Music*, Camden, NJ, 1913.

Feaster, Patrick, 'Framing the Mechanical Voice: Generic conventions of early phonograph recording', *Folklore Forum*, 32, 1 & 2, 2001, pp. 57–102.

Feaster, Patrick and Richard Bauman, '"Fellow Townsman and My Noble Constituents": Representations of oratory on early commercial recordings', *Oral Tradition*, 20, 1, 2005, pp. 35–57.

Feaster, Patrick and Richard Bauman, 'Oratorical Footing in a New Medium: Recordings of presidential campaign speeches 1896–1912', *Proceedings of the Eleventh Annual Symposium About Language and Society, Austin, April 11–13, 2003*, Austin, 2004, pp. 1–19.

Feaster, Patrick and Jacob Smith, 'Reconfiguring the History of Early Cinema Through the Phonograph, 1877–1908', *Film History*, 21, 4, 2009, pp. 311–25.

Findlay, Mary, *Tooth and Nail: The story of a daughter of the Depression*, Auckland, 1989.

Fischer, Claude, *America Calling: A social history of the telephone to 1940*, Berkeley, CA, 1992.

Fischer, Claude, 'Gender and the Residential Telephone 1890–1940: Technologies of sociability', *Sociological Forum*, 3, 2, 1988, pp. 211–33.

Fitzgerald, A.N., 'New Zealand Speech: Impressions and reflections', *Education Gazette*, 13, 1934, p. 123.

Flanders, Judith, *Consuming Passions: Leisure and pleasure in Victorian England*, London, 2006.

Flanders, Judith, *The Victorian House: Domestic life from childbirth to deathbed*, London, 2003.

Foote, Chris, 'Broadcasting in New Zealand and the Role of the Post Office: Part One', *The Mail Coach: Journal of the Postal History Society of New Zealand*, 47, 2, 2010, pp. 66–80.

Foote, Chris, 'Broadcasting in New Zealand and the Role of the Post Office: Part Two', *The Mail Coach: Journal of the Postal History Society of New Zealand*, 48, 3, 2012, pp. 104–19.

Foote, Chris, 'Broadcasting in New Zealand and the Role of the Post Office: Part Three', *The Mail Coach: Journal of the Postal History Society of New Zealand*, 48, 5, 2012, pp. 186–95.

Forest, Jennifer, 'Scripting the Female Voice: The phonograph, the cinematograph, and the ideal woman', *Nineteenth-Century French Studies*, 17, 1 & 2, 1998–99, pp. 71–95.
Fowler, Gene and Bill Crawford, *Border Radio: Quacks, yodelers, pitchmen, psychics, and other amazing broadcasters of the American airwaves*, Austin, 2002.
Fowler, Michael and David Turnbull, *The Reel Story: A history of Napier and Hastings cinemas 1896–1996*, Havelock North, 2008.
Fox, Alistair, 'Rudall Hayward and the Cinema of Maoriland: Genre-mixing and counter-discourses in *Rewi's Last Stand* (1925), *The Te Kooti Trail* (1927) and *Rewi's Last Stand/The Last Stand* (1940)', in Alistair Fox, Barry Keith Grant and Hilary Radner, eds, *New Zealand Cinema: Interpreting the past*, Bristol, 2011, pp. 45–64.
Fox, Alistair, Barry Keith Grant and Hilary Radner, eds, *New Zealand Cinema: Interpreting the past*, Bristol, 2011.
Francis, Bill, *ZB: The voice of an iconic station*, Auckland, 2006.
Frattarola, Angela, 'Developing an Ear for the Modernist Novel: Virginia Woolf, Dorothy Richardson and James Joyce', *Journal of Modern Literature*, 33, 1, 2009, pp. 132–53.
Freeman, Graham, 'That Chief Undercurrent of My Mind: Percy Grainger and the aesthetics of English folk song', *Folk Music Journal*, 9, 4, 2009, pp. 581–617.
Freire, Sérgio, 'Early Musical Impressions from Both Sides of the Loudspeaker', *Leonardo Music Journal*, 13, 2003, pp. 67–71.
Frith, Simon, 'Art Versus Technology: The strange case of popular music', *Media, Culture and Society*, 8, 3, 1986, pp. 263–79.
Frith, Simon, *Performing Rites: On the value of popular music*, Cambridge, MA, 1996.
Froude, Tony, *Big Screens in the Valley: Cinemas and movie halls in the Hutt Valley 1906–2002*, Paraparaumu, 2002.
Froude, Tony, *Where To Go on Saturday Nights: Wellington cinemas and movie halls 1896–2000*, Wellington, 2001.
Fry, A.S., *The Aunt Daisy Story*, Wellington, 1957.
Fry, Margot, ed., *A Century of Film in New Zealand: Papers from the conference 'Cinema, Film & Society', Wellington, October, 1996*, Wellington, 1998.
Fuller, Sarah, 'Delectabatur In Hoc Auris: Some fourteenth-century perspectives on aural perception', *The Musical Quarterly*, 82, 3/4, 1998, pp. 466–81.
Fyfe, Judith, *Matriarchs: A generation of New Zealand women talk to Judith Fyfe*, Auckland, 1990.
Gaisberg, Fred, *Music on Record*, London, 1946.
Garofalo, Reebee, 'From Music Publishing to MP3: Music and industry in the twentieth century', *American Music*, 17, 3, 1999, pp. 318–54.
Garrioch, David, 'Sounds of the City: The soundscape of early modern European towns', *Urban History*, 30, 1, 2003, pp. 5–25.
Gaughie, Pamela L., ed., *Virginia Woolf in the Age of Mechanical Production*, New York, 2000.
Gelatt, Roland, *The Fabulous Phonograph 1877–1977*, 2nd edn, London, 1977.
Gerstner, David and Sarah Greenlees, 'Cinema by Fits and Starts: New Zealand film practices in the twentieth century', *Cineaction*, 51, 2000, pp. 46–57.

Gibbons, Peter, 'Cultural Colonization and National Identity', *New Zealand Journal of History*, 36, 1, 2002, pp. 5–17.
Gibbons, Peter, 'The Far Side of the Search for Identity: Reconsidering New Zealand history', *New Zealand Journal of History*, 37, 1, 2003, pp. 38–49.
Gibbons, Peter, 'Non-Fiction', in Terry Sturm, ed., *The Oxford History of New Zealand Literature in English*, 2nd edn, Oxford, 1998, pp. 31–118.
Giddens, Anthony, *Consequences of Modernity*, Stanford, 1991.
Giddens, Anthony, *Modernity and Self- Identity: Self and society in the late modern age*, Cambridge, 1991.
Gitelman, Lisa, 'How Users Define New Media: A history of the amusement phonograph', in David Thorburn and Henry Jenkins, eds, *Rethinking Media Change: The aesthetics of transition*, Cambridge, MA, 2003, pp. 61–80.
Gitelman, Lisa, 'Reading Music, Reading Records, Reading Race: Musical copyright and the U.S. Copyright Act of 1909', *The Musical Quarterly*, 81, 2, 1997, pp. 265–90.
Gitelman, Lisa, *Scripts, Grooves and Writing Machines: Representing technology in the Edison era*, Stanford, 1999.
Gitelman, Lisa, 'Souvenir Foils: On the status of print at the origin of recorded sound', in Lisa Gitelman and Geoffrey Rice, eds, *New Media 1740-1915*, Cambridge, MA, 2003, pp. 157–73.
Gitelman, Lisa, 'Unexpected Pleasures: Phonographs and cultural identities in America 1895–1915', in Ron Eglash, Jennifer L. Croissant, Giovanna Di Chiro, Rayvon Fouché et al., eds, *Appropriating Technology: Vernacular science and social power*, Minneapolis, 2004, pp. 331–44.
Gitelman, Lisa and Geoffrey Rice, eds, *New Media 1740-1915*, Cambridge, MA, 2003.
Goldmark, Daniel, Lawrence Kramer and Richard Leppert, eds, *Beyond the Soundtrack: Representing music in cinema*, Berkeley, CA, 2007.
Golledge, Wally, 'Early Days of the Phonograph in Nelson', *The Phonographic Record: Supplement*, 4, 5, 1969, pp. 1–3.
Gomery, Douglas, *The Coming of Sound: A history*, New York, 2005.
Good, Katie Day, 'Radio's Forgotten Visuals', *Journal of Radio and Audio Media*, 23, 2, 2016, pp. 364–68.
Goodale, Greg, *Sonic Persuasion: Reading sound in the recorded age*, Chicago, 2011.
Goodman, Dave, 'Distracted Listening: On not making sound choices in the 1930s', in David Suisman, David and Susan Strasser, eds, *Sound in the Age of Mechanical Reproduction*, Philadelphia, 2010, pp. 15–46.
Gouk, Penelope, *Music, Science and Natural Magic in Seventeenth-century England*, New Haven, 1999.
Gracyk, Theodore, 'Listening to Music: Performances and recordings', *The Journal of Aesthetics and Art Criticism*, 55, 2, 1997, pp. 139–50.
Gracyk, Tim and Frank Hoffmann, *Popular American Recording Pioneers: 1895-1925*, New York, 2000.
Grajeda, Tony and Jay Beck. 'Introduction: The future of Sound Studies', *Sound, and the Moving Image*, 2, 2, 2008, pp. 109–14.
The Gramophone Company, *Opera At Home*, London, 1920.

Greenhalgh, Charlotte, 'Bush Cinderellas: Young New Zealanders and romance at the movies 1919–1939', *New Zealand Journal of History*, 44, 1, 2010, pp. 1–21.

Grieveson, Lee and Peter Krämer, eds, *The Silent Cinema Reader*, Oxford, 2006.

Griffen-Foley, Bridget, 'Modernity, Intimacy and Early Australian Commercial Radio', in Joy Damousi and Desley Deacon, eds, *Talking and Listening in the Age of Modernity: Essays on the history of sound*, Canberra, 2007, pp. 123–32.

Griffiths, John, 'Popular Culture and Modernity: Dancing in New Zealand society 1920–1945', *Journal of Social History*, 41, 13, 2008, pp. 611–32.

Gronow, Pekka, 'The Record Industry Comes to the Orient', *Ethnomusciology*, 25, 2, 1981, pp. 251–84.

Gronow, Pekka, 'The Record Industry: Growth of a mass culture', *Popular Music*, 3, 1983, pp. 53–75.

Gronow, Pekka and Björn Englund, 'Inventing Recorded Music: The recorded repertoire in Scandinavia 1899–1925', *Popular Music*, 26, 2, 2007, pp. 281–304.

Gross, John, ed., *The Oxford Book of Comic Verse*, Oxford, 1994.

Gue, Randy, '"It Seems that Everything Looks Good Nowadays, as Long as it is in the Flesh & Brownskin": The assertion of cultural difference at Atlanta's 81 Theatre, 1934–1937', *Film History*, 8, 2, 1996, pp. 209–18.

Gunning, Tom, 'Doing for the Eye What the Phonograph Does for the Ear', in Richard Abel and Rick Altman, eds, *The Sounds of Early Cinema*, Bloomington, 2001, pp. 13–31.

Gunning, Tom, 'Now You See It, Now You Don't: The temporality of the cinema of attractions', in Lee Grieveson and Peter Krämer, eds, *The Silent Cinema Reader*, Oxford, 2006, pp. 41–50.

Haddon, Kathy, *Birkenhead: The way we were*, Birkenhead, 1993.

Hall, John, *The History of Broadcasting in New Zealand 1920–1954*, Wellington, 1984.

Hardt, Hanno, *In the Company of Media: Cultural constructions of communication 1920s–1930s*, Boulder, CO, 2000.

Haring, Kristen, 'The "Freer Men" of Ham Radio: How a technical hobby provided social and spatial distance', *Technology and Culture*, 44, 4, 2003, pp. 734–61.

Harper, Glyn, *Johnny Enzed: The New Zealand soldier in the First World War 1914–1918*, Auckland, 2015.

Harrison, P.A., 'The Motion Picture Industry in New Zealand 1896-1930: A history of the commercial distribution and exhibition of films', MA thesis, University of Auckland, 1974.

Hartrick, Elizabeth, 'Consuming Illusions: The magic lantern in Australia and Aotearoa/New Zealand 1850–1910', PhD thesis, The Australian Centre, University of Melbourne, 2003.

Hawkey, Rachel, 'A Man of Missionary Zeal: Vernon Griffiths', *Sound Ideas*, 6, 1, 2003, pp. 19–25.

Hawkey Rachel, 'This is a British Colony: Vernon Griffiths & music education in New Zealand', *Music in New Zealand*, 14, 1991, pp. 30–36.

Hawkey, Rachel, 'Vernon Griffiths: His life and philosophy of music education as demonstrated through his collected papers', PhD thesis, Canterbury University, 1993.

Haworth, Diane, *Father & Son: The Bob Kerridge story*, Auckland, 2004.

Hayward, Bruce W. and Selwyn p. Hayward, *Cinemas of Auckland 1896–1979*, Auckland, 1979.

Hayward, Henry J., *Here's to Life: The impressions, confessions and garnered thoughts of a free-minded showman*, Auckland, 1944.

Hegarty, Paul, *Noise/Music: A history*, New York, 2007.

Heilbroner, R.L. 'Do Machines Make History?', in Merritt Roe Smith and Leo Marx, eds, *Does Technology Drive History?: The dilemma of technological determinism*, Cambridge, MA, 1998, pp. 53–65.

Hendrykowska, Malgorzata, 'From the Phonograph to the Kinetophone: Sound in the cinema within the Polish territory prior to 1914', *Film History*, 11, 4, 1999, pp. 444–48.

Herriman, Nina, 'The Air Down Here: Global and local interpretations of New Zealand popular music, 1955–1977', MA thesis, University of Auckland, 2004.

High, Peter, 'The Dawn of Cinema in Japan', *Journal of Contemporary History*, 19, 1, 1984, pp. 23–57.

Hill, Jonathan, *The Cat's Whisker: 50 Years of wireless design*, London, 1978.

Hilliard, Chris, *The Bookmen's Dominion: Cultural life in New Zealand 1920–1950*, Auckland, 2006.

Hillyer, Minette, 'Dominion Screen Types and Local Beauty Spots: New Zealand's pre-war community films', *New Zealand Journal of Media Studies*, 11, 2, 2008, pp. 37–57.

Hilmes, Michele, 'Foregrounding Sound: New (and old) directions in Sound Studies', *Cinema Journal*, 48, 1, 2008, pp. 115–17.

Hilmes, Michele, 'Is There a Field Called Sound Studies? And does it matter?', *American Quarterly*, 57, 1, 2005, pp. 249–59.

Hilmes, Michele, *Radio Voices: American broadcasting 1922–1952*, Minneapolis, 1997.

Hilmes, Michele and Jason Loviglio, eds, *Radio Reader: Essays in the cultural history of radio*, New York, 2002.

Hirt, Katherine, *When Machines Play Chopin: Musical spirit and automation in nineteenth-century German literature*, Berlin, 2010.

Hjorth, Larissa, Jean Burgess and Ingrid Richardson, eds, *Studying Mobile Media: Cultural technologies, mobile communication and the iPhone*, New York, 2012.

Hoar, Peter, 'A Qualitative Content Analysis of the New Zealand Troopship Publications 1914–1920', MLIS thesis, Victoria University of Wellington, 2001.

Hohepa, Pat, 'Maori and Pakeha: The one-people myth', in Michael King, ed., *Tihe Mauri Ora: Aspects of Maoritanga*, Auckland, 1978, pp. 98–111.

Hoorn, Jeanette and Michelle Smith, 'Rudall Hayward's Democratic Cinema and the "Civilising Mission" in the "Land of the Wrong White Crowd"', in Alistair Fox, Barry Keith Grant and Hilary Radner, eds, *New Zealand Cinema: Interpreting the past*, Bristol, 2011, pp. 65–81.

Hosokawa, Shuhei, 'The Walkman Effect', *Popular Music*, 4, 1984, pp. 165–80.

Howes, David, ed., *Empire of the Senses: The sensual culture reader*, Oxford, 2005.

Hunt, A.N., 'Recreation and Entertainment', in A.N. Hunt, ed., *Foxton 1888–1988: The first 100 years*, Foxton, 1987.

Hurst, Maurice, *Music and the Stage in New Zealand: The record of a century of entertainment*, Auckland, 1944.
Hyde, Robyn, *Journalese*, Wellington, 1934.
Hyde, Robin, *Disputed Ground: Robin Hyde journalist*, Wellington, 1991.
Ihde, Don, *Listening and Voice: Phenomenologies of sound*, 2nd edn, New York, 2007.
Ingham, Gordon, *Everybody's Gone to the Movies: The sixty cinemas of Auckland and some others*, Auckland, 1973.
Jalland, Patricia, *Death in the Victorian Family*, New York, 1996.
Jay, Martin, *Downcast Eyes: The denigration of vision in twentieth-century French thought*, Berkeley, CA, 1993.
Jennings, John M., *Music at Canterbury: A centennial history of the School of Music, University of Canterbury, Christchurch, New Zealand, 1891–1991*, Christchurch, 1991.
Jensen, Peter R., *In Marconi's Footsteps: Early radio 1894 to 1920*, Kenthurst, NSW, 1994.
Jewell, Brian, *Veteran Talking Machines*, Tunbridge Wells, 1977.
Johnson, Bruce, *The Inaudible Music: Jazz, gender and Australian modernity*, Sydney, 2000.
Johnson, Harry, ed., *Many Voices: Music and national identity in Aotearoa/New Zealand*, Newcastle, 2010.
Johnson, James H., *Listening in Paris: A cultural history*, Berkeley, CA, 1995.
Johnson, Lesley, *The Unseen Voice: A cultural study of early Australian radio*, London, 1988.
Johnson, William Ward, *The Gramophone in Education: An introduction to its use in school and in the home*, London, 1936.
Johnson, William Ward, *Intelligent Listening to Music: A guide to enjoyment and appreciation for all lovers of music*, 2nd edn, London, 1939.
Jones, Andrew F., *Yellow Music: Media culture and colonial modernity in the Chinese jazz age*, Durham, NC, 2001.
Jones, Geoffrey, 'The Gramophone Company: An Anglo-American multinational 1898–1931', *Business History Review*, 59, 1, 1985, pp. 76–100.
Joyce, James, *Ulysses*, Oxford, 1993.
Kahn, Douglas, *Noise Water Meat: A history of sound in the arts*, Cambridge, MA, 1999.
Kahn, Douglas and Gregory Whitehead, eds, *Wireless Imagination: Sound, radio, and the avant-garde*, Cambridge, MA, 1992.
Katz, Mark, *Capturing Sound: How technology has changed music*, Berkeley, CA, 2004.
Keam, Glenda and Tony Mitchell, eds, *Home, Land and Sea: Situating music in Aotearoa New Zealand*, Auckland, 2011.
Keane, Damien, 'Quotation Marks, the Gramophone and the Language of the Outlaw', *Texas Studies in Literature and Language*, 51, 4, 2009, pp. 400–15.
Keeling, Kara and Josh Kun, 'Introduction: Listening to American Studies', *American Quarterly Special Issue: Listening to American Studies*, 63, 3, 2011, pp. 445–59.
Keightley, Emily, ed., *Time, Media and Modernity*, London, 2012.
Keightley, Emily and Michael Pickering, 'For the Record: Popular music and photography as technologies of memory', *European Journal of Cultural Studies*, 9, 2, 2006, pp. 149–65.

Kenney, William H., *Recorded Music in American Life: The phonograph and popular memory 1890–1945*, New York, 1999.
Kenyon, N., ed., *Authenticity and Early Music: A symposium*, Oxford, 1988.
Kerins, Mark, 'A Statement on Sound Studies (with apologies to Sergei Eisenstein, Vsevolod Pudovkin, and Grigori Alexandrov)', *Sound and the Moving Image*, 2, 2, 2008, pp. 115–19.
Kerr, Donald, *Amassing Treasures for All Times: Sir George Grey, colonial bookman and collector*, Dunedin, 2006.
Kern, Stephen, *The Culture of Time and Space 1880–1918*, London, 1983.
Kierkegaard, Søren, *Either/Or Part II*, Princeton, 1987.
King, Michael, *The Penguin History of New Zealand*, Auckland, 2003.
King, Michael, ed., *Tihe Mauri Ora: Aspects of Maoritanga*, Auckland, 1978.
Kittler, Friedrich A., *Discourse Networks 1800/1900*, Stanford, 1990.
Kittler, Friedrich A., *Gramophone, Film, Typewriter*, Stanford, 1999.
Klein, Herman, 'Sims Reeves: Prince of English tenors', in Roger Wimbush, ed., *The Gramophone Jubilee Book 1923–1973*, Harrow, 1973, pp. 109–12.
Knewstubb Theatres, *Cinemas: Dunedin and districts 1897–1974*, Dunedin, 1974.
Knowles, Sebastian D.G., 'Death by Gramophone', *Journal of Modern Literature*, 27, 1 & 2, 2003, pp. 1–13.
Kraft, James P., *Stage to Studio: Musicians and the sound revolution 1890–1950*, Baltimore, 1996.
Kreilkamp, Ivan, 'A Voice Without a Body: The phonographic logic of "Heart of Darkness"', *Victorian Studies*, 40, 2, 1997, pp. 211–44.
Kruse, Holly, 'Early Audio Technology and Domestic Space', *Stanford Humanities Review*, 3, 2, 1993, pp. 1–14.
Lacey, Kate, 'Towards a Periodization of Listening: Radio and modern life', *International Journal of Cultural Studies*, 3, 2, 2000, pp. 279–88.
Laing, Dave, 'A Voice without a Face: Popular music and the phonograph in the 1890s', *Popular Music*, 10, 1, 1991, pp. 1–9.
Laird, Ross, *Sound Beginnings: The early record industry in Australia*, Sydney, 1999.
Lambert, Constant, *Music Ho! A Study of Music in Decline*, London, 1934.
Lascelles, David, *Eighty Turbulent Years: The Paramount Theatre Wellington 1917–1997*, Wellington, 1997.
Lash, Scott and Jonathan Friedman, 'Introduction: Subjectivity and modernity's Other', in Scott Lash and Jonathan Friedman, eds, *Modernity and Identity*, Oxford, 1992, pp. 1–30.
Lash, Scott and Jonathan Friedman, eds, *Modernity and Identity*, Oxford, 1992.
Lastra, James, *Sound Technology and the American Cinema: Perception, representation, modernity*, New York, 2000.
Latchem, Jane, 'A Generation of Promise: The 1908 Junior National Scholarship candidates – education, occupation, and the First World War', *Journal of New Zealand Studies*, 18, 2014, pp. 66–84.
Laurenson, Helen B., *Going Up, Going Down: The rise and fall of the department store*, Auckland, 2005.
Lealand, Geoff, *A Foreign Egg in Our Nest?: American popular culture in New Zealand*, Wellington, 1988.

Lee, John A., *Early Days in New Zealand*, Martinborough, 1977.
Leech-Wilkinson, Daniel, 'Recordings and Histories of Performance Style', in Nicholas Cook, Eric Clarke, Daniel Leech-Wilkinson and John Rink, eds, *The Cambridge Companion to Recorded Music*, Cambridge, 2009, pp. 246–62.
LeMahieu, Don, *A Culture for Democracy: Mass communication and the cultivated mind in Britain between the wars*, Oxford, 1988.
LeMahieu, Don, 'The Gramophone: Recorded music and the cultivated mind in Britain between the wars', *Technology and Culture*, 23, 3, 1982, pp. 372–91.
Lemke, Claudia, 'Maori Involvement in Sound Recording and Broadcasting 1919 to 1958', MA thesis, University of Auckland, 1995.
Leppert, Richard and Susan McClary, eds, *Music and Society: The politics of composition, performance and reception*, Cambridge, 1987.
Levin, David, ed., *Modernity and the Hegemony of Vision*, Berkeley, CA, 1993.
Levine, Lawrence, *Highbrow/Lowbrow: The emergence of cultural hierarchy in America*, Cambridge, MA, 1988.
Levine, Lawrence, 'Jazz and American Culture', *The Journal of American Folklore*, 102, 403, 1989, pp. 6–22.
Levitin, Daniel J., *This is Your Brain on Music: The science of a human obsession*, New York, 2006.
Levy, Steven, *The Perfect Thing: How the iPod shuffles commerce, culture and coolness*, New York, 2006.
Lewis, Laurie, *Arthur and the Nights at the Turntable: The life and times of a jazz broadcaster*, London, 1996.
Link, Stan, 'The Work of Reproduction in the Mechanical Aging of Art: Listening to noise', *Computer Music Journal*, 25, 1, 2001, pp. 34–47.
Locke, Ralph P., 'Music Lovers, Patrons and the "Sacralization" of Culture in America', *19th-Century Music*, 17, 2, 1993, pp. 149–73.
Lockheart, Paula, 'A History of Early Microphone Singing 1925–1939: American mainstream popular singing at the advent of electronic microphone amplification', *Popular Music and Society*, 26, 3, 2003, pp. 367–85.
Lodge, Oliver, *Signalling Through Space Without Wires*, London, 1900.
Lopes, Paul, 'Diffusion and Syncretism: The modern jazz tradition', *Annals of the American Academy of Political and Social Science*, 566, 1999, pp. 25–36.
'Lorenski', 'Shut That Old Piano Up: Written for the new phonograph', *The New Phonogram*, 2, 2, 1910, pp. 10–11.
Luce, R.D., *Sound and Hearing: A conceptual introduction*, Hilldale, 1993.
Macdonald, Charlotte, Merimeri Penfold and Bridget Williams, eds, *The Book of New Zealand Women: Ko kui ma te Kaupapa*, Wellington, 1991.
McDouall, Hamish, *100 Essential New Zealand Films*, Wellington, 2009.
MacGibbon, John, *Piano in the Parlour: When the piano was New Zealand's home entertainment centre*, Wellington, 2007.
Mackay, Ian K., *Broadcasting in New Zealand*, Wellington, 1953.
Mackenzie, Compton, 'The Gramophone: Its past, its present, its future', *Proceedings of the Musical Association*, 51st Sess., London, 1924–1925, pp. 97–119.
Mackenzie, Compton, *A Musical Chair*, London, 1939.

Mackenzie, Compton, *My Record of Music*, London, 1955.
Main, William, '"The Lanthorn that Shews Tricks": The magic lantern in nineteenth-century New Zealand', *Turnbull Library Record*, 27, 1994, pp. 45–53.
Makagon, Daniel and Mark Neumann, *Recording Culture: Audio documentary and the ethnographic experience*, Los Angeles, 2009.
Marks, Martin Miller, *Music and the Silent Film: Contexts and case studies 1895–1924*, Oxford, 1997.
Martin, Helen and Sam Edwards, *New Zealand Film 1912–1996*, Auckland, 1997.
Marvin, Carolyn, *When Old Technologies Were New: Thinking about electric communication in the late nineteenth century*, Oxford, 1988.
Masters, Alison, 'Ana Hato 1906–1953', in Charlotte Macdonald, Merimeri Penfold and Bridget Williams, eds, *The Book of New Zealand Women: Ko kui ma te Kaupapa*, Wellington, 1991, pp. 277–78.
Mathieson, Eileen, 'The Gaiety Picture Theatre: Paeroa', *Ohinemuri Regional History Journal*, 32, 1988, p. 48.
Matthews, Jill Julius, *Dance Hall & Picture Palace: Sydney's romance with modernity*, Sydney, 2005.
McCracken, Alison, 'God's Gift to Us Girls: Crooning, gender and the re-creation of American popular song 1928–1933', *American Music*, 17, 4, 1999, pp. 365–95.
McCracken, Allison, 'Real Men Don't Sing Ballads: The radio crooner in Hollywood', in Pamela Robertson Wojcik and Arthur Knight, eds, *Soundtrack Available: Essays on film and popular music*, Durham, NC, 2001, pp. 105–34.
McGibbon, Ian, ed., *The Oxford Companion to New Zealand Military History*, Auckland 2000.
McKinlay, Ernest, *Ways and By-Ways of a Singing Kiwi*, Dunedin, 1939.
McKinnon, Malcolm, ed., *The American Connection: Essays from the Stout Centre Conference*, Wellington, 1988.
McLuhan, Marshall, *The Gutenberg Galaxy: The making of typographic man*, Toronto, 1962.
McLuhan, Marshall, *Understanding Media: The extensions of man*, London, 1964.
McMahon, Morgan, *Vintage Radio: A pictorial history of wireless and radio 1887–1929*, 3rd edn, Palos Verdes Peninsula, CA, 1981.
Melosi, Martin V., ed., *Pollution and Reform in American Cities 1879–1930*, Austin, 1980.
Melville-Mason, Graham, 'The Gramophone as Furniture', in Alistair G. Thomson, ed., *Phonographs & Gramophones: A symposium organised by the Royal Scottish Museum in connection with the exhibition Phonographs and Gramophones and the Centenary of the Invention of the Phonograph by Thomas Alva Edison*, Edinburgh, 1977, pp. 117–38.
Mercer, David, *The Telephone: The life story of a technology*, Westport, CT, 2006, pp. 39–56.
Meszaros, Beth, 'Infernal Sound Cues: Aural geographies and the politics of noise', *Modern Drama*, 48, 1, 2005, pp. 118–31.
Middleton, Richard, 'In the Groove or Blowing Your Mind?: The pleasures of musical repetition', in Andy Bennett, Barry Shank and Jason Toynbee, eds, *The Popular Music Studies Reader*, Oxford, 2000, pp. 15–20.

Millard, Andre, *America on Record: A history of recorded sound*, 2nd edn, New York, 2005.
Miller, Russell and Roger Boar, *The Incredible Music Machine*, London, 1982.
Milner, Greg, *Perfecting Sound Forever: The story of recorded music*, London, 2009.
Mirams, Gordon, *Speaking Candidly: Films and people in New Zealand*, Hamilton, 1945.
Moffat, Kirstine, *Piano Forte: Stories and soundscapes from colonial New Zealand*, Dunedin, 2011.
Misa, Thomas J., 'The Compelling Tangle of Modernity and Technology', in Thomas, J. Misa, Philip Brey and Andrew Feenberg, eds, *Modernity and Technology*, Cambridge, MA, 2003, pp. 1–30.
Misa, Thomas, J. Philip Brey and Andrew Feenberg, eds, *Modernity and Technology*, Cambridge, MA, 2003.
Monckton, C.C.F. *Radio-Telegraphy*, London, 1908.
Montgomerie, Deborah, *Love in Time of War: Letter writing in the Second World War*, Auckland, 2005.
Moore, Jerrold Northrop, *Sound Revolutions: A biography of Fred Gaisberg, founding father of commercial sound recording*, 2nd edn, London, 1999.
Moores, Shaun, '"The Box on the Dresser": Memories of early radio and everyday life', *Media, Culture and Society*, 10, 1, 1988, pp. 23–40.
Morrison, Simon, 'Skryabin and the Impossible', *Journal of the American Musicological Society*, 51, 2, 1998, pp. 283–330.
Morton, David, *Off the Record: The technology and culture of sound recording in America*, New Brunswick, NJ, 2000.
Moresby, T.A., 'Music in Paeroa from 1908 to 1928', *Ohinemuri Regional History Journal*, 6, 1, 1969, pp. 18–21.
Mowitt, John, 'The Sound of Music in the Era of its Electronic Reproducibility', in Richard Leppert and Susan McClary, eds, *Music and Society: The politics of composition, performance and reception*, Cambridge, 1987, pp. 173–97.
Mulgan, Alan, *The Making of a New Zealander*, Wellington, 1958.
Murray, David, 'Fitchett's Fallacy and Music at the New Zealand and South Seas Exhibition, Dunedin, 1889–1890', *New Zealand Journal of History*, 42, 1, 2008, pp. 42–59.
Musser, Charles, *The Emergence of Cinema: The American screen to 1907*, Berkeley, CA, 1990.
Musser, Charles, 'Moving Towards Fictional Narratives: Story films become the dominant product 1903–1904', in Lee Grieveson and Peter Krämer, eds, *The Silent Cinema Reader*, Oxford, 2006, pp. 87–102.
Naeem, Asma, 'Splitting Sight and Sound: Thomas Dewing's *A Reading*, gilded age women and the phonograph', *American Quarterly*, 63, 3, 2011, pp. 461–85.
Nancy, Jean-Luc, *Listening*, New York, 2007.
The National Phonograph Company, *The Phonograph and How to Use It*, 1900, New York, facsimile, 1900.
Navitski, Rielle, 'The Tango on Broadway: Carlos Gardel's international stardom and the transition to sound in Argentina', *Cinema Journal*, 51, 1, 2011, pp. 26–49.

Negus, Keith, *Music Genres and Corporate Cultures*, London, 1999.

Negus, Keith, *Producing Pop: Culture and conflict in the recording industry*, New York, 2001.

Neuman, Dard, 'The Production of Aura in the Gramophone Age of the "Live" Performance', *Asian Music*, 40, 2, 2009, pp. 101–23.

Nickles, Shelley, 'Object Lessons: Household appliance design and the American middle class, 1920–1960', PhD thesis, University of Virginia, 1999.

Nicolson, Malcolm, 'The Introduction of Percussion and Stethoscopy to Early Nineteenth Century Edinburgh', in W.F. Bynum and Roy Porter, eds, *Medicine and the Five Senses*, Cambridge, 1993, pp. 134–53.

Nobbs, George, *The Wireless Stars*, Norwich, 1972.

Norman, Philip, *Douglas Lilburn: His life and music*, Christchurch, 2006.

Norris, Walter, 'Ana Hato and Deane Waretini', *The Phonographic Record*, 2, 5, 1967, pp. 36–37.

Norris, Walter, 'The Edison Phonograph in Christchurch between 1879 and 1914', *The Phonographic Record*, 8, 4, 1973, pp. 79–80.

Nott, James J., *Music for the People: Popular music and dance in interwar Britain*, Oxford, 2002.

Nunley, C., 'For the Record: Robert Desnos, music, and wartime memory', *SubStance*, 38, 2, 2009, pp. 113–35.

O'Callaghan, Casey, *Sounds: A philosophical theory*, Oxford, 2007.

O'Callahan, Alice, 'I Remember', in Alec Utting, ed., *Life, Laughter and Love in the Early Years: Remembering earlier times in Birkenhead and Birkdale*, Birkenhead, 2004, pp. 39–41.

Oja, Carol J., 'Gershwin and American Modernists of the 1920s', *The Musical Quarterly*, 78, 4, 1994, pp. 646–68.

Olsson, Jan, 'In and Out of Sync: Swedish sound films 1903–1914', *Film History*, 11, 4, 1999, pp. 449–55.

Openshaw, Roger and Roy Shuker, 'Silent Movies and Comics', in Malcolm McKinnon, ed., *The American Connection: Essays from the Stout Centre Conference*, Wellington, 1988, pp. 52–65.

O'Rawe, Des, 'The Great Secret: Silence, cinema and modernism', *Screen*, 47, 4, 2006, pp. 395–405.

O'Shea, John, 'A Charmed Life: Fragments of memory', in Jonathan Dennis and Jan Bieringa, eds, *Film in Aotearoa New Zealand*, Wellington, 1992, pp. 13–35.

O'Sullivan, Vincent and Margaret Scott, eds, *The Collected Letters of Katherine Mansfield*, Oxford, 1984.

Oswell, David, 'Early Children's Broadcasting in Britain: Programming for a liberal democracy', *Historical Journal of Film, Radio and Television*, 18, 3, 1998, pp. 375–93.

Owen, Graham, 'Expectations, Panic and Change: The early years of motion pictures in New Zealand', in Margot Fry, ed., *A Century of Film in New Zealand: Papers from the conference 'Cinema, Film & Society'*, Wellington, October, 1996, Wellington, 1998, pp. 20–27.

Parakilas, James, ed., *Piano Roles: A new history of the piano*, 2nd ed., New Haven, 2002.

Parekowhai, Michael, *Ten Guitars*, Auckland, 1999.
Park, David W., Nicholas W. Jankowski and Steve Jones, eds, *The Long History of New Media: Technology, historiography, and contextualising newness*, New York, 2011.
Pawson, Eric and Neil C. Quigley, 'The Circulation of Information and Frontier Development: Canterbury 1850–1890', *New Zealand Geographer*, 38, 2, 1982, pp. 65–76.
Peters, John Durham, *Speaking into the Air: A history of the idea of communication*, Chicago, 1999.
Peterson, Bert, 'Epi Shalfoon: Loss of popular musician', *Te Ao Hou*, 5, 1953, pp. 19–20.
Petts, Leonard, *The Story of 'Nipper' and the 'His Master's Voice' Painted by Francis Barraud*, London, 1973.
Philip, Robert, *Performing Music in the Age of Recording*, New Haven, 2004.
Phillips, Jock, 'Exhibiting Ourselves: The exhibition and national identity', in J.M. Thomson, ed., *Farewell Colonialism: The New Zealand Exhibition Christchurch 1906–1907*, Palmerston North, 1998, pp. 17–26.
Phillips, Jock, *A Man's Country? The image of the Pakeha male: A history*, Auckland, 1987.
Phillips, Jock, 'Men, Women and Leisure since the Second World War', in Caroline Daley and Deborah Montgomerie, eds, *The Gendered Kiwi*, Auckland, 1999, pp. 213–33.
Phillips, Jock, 'Our History, Our Selves: The historian and national identity', *New Zealand Journal of History*, 30, 2, 1996, pp. 107–23.
Phillips, Ray, *Edison's Kinetoscope and its Films: A history to 1896*, Westport, CT, 1997.
Picker, John, 'The Soundproof Study: Victorian professionals, work space and urban noise', *Victorian Studies*, 42, 3, 1999, pp. 427–53.
Picker, John, *Victorian Soundscapes*, Oxford, 2003.
Pickering, Michael, 'Sonic Horizons: Phonograph aesthetics and the experience of time', in Emily Keightley, ed., *Time, Media and Modernity*, London, 2012, pp. 25–44.
Pierce, George, *Principles of Wireless Telegraphy*, New York, 1910.
Pinch, Trevor and Karin Bijsterveld, 'Sound Studies: New technologies and music', *Social Studies of Science*, 34, 5, 2004, pp. 635–48.
Pivac, Diane, 'The Rise of Fiction: Between the wars', in Diane Pivac, Frank Stark and Lawrence McDonald, eds, *New Zealand Film: An illustrated history*, pp. 68–69.
Pivac, Diane and Robert Sklar, 'Rudall Hayward, New Zealand Film-Maker', *Landfall*, 98, 1971, pp. 147–54.
Pivac, Diane, Frank Stark and Lawrence McDonald, eds, *New Zealand Film: An illustrated history*, Wellington, 2011.
Porter, Roy, ed., *Rewriting the Self: Histories from the Renaissance to the present*, London, 1997.
Potter, Simon J., *Broadcasting Empire: The BBC and the British world 1922–1970*, Oxford, 2012.
Powles, C.G., H.G. Alexander and H.A. Lockington, 'N.Z. Wireless Troop in Mesopotamia and Persia', in N. Annabell, ed., *Official History of the New*

Zealand Engineers during the Great War, 1914–1919: A record of the work carried out by the field companies, field troops, Signal Troop and Wireless Troop, during the operations in Samoa (1914–15); Egypt, Gallipoli, Sinai and Palestine (1914–18); France, Belgium and Germany (1916–19); and Mesopotamia (1916–18)*, Wanganui, 1927, pp. 297–303.

Price, Simon, *New Zealand's First Talkies: Early film-making in Otago and Southland 1896–1939*, Dunedin, 1996.

Pridmore-Brown, Michele, '1939–40: Of Virginia Woolf, gramophones and fascism', *PMLA: Publications of the Modern Language Association of America*, 113, 3, 1998, pp. 408–21.

Priestley, Brian, *Jazz on Record: A history*, London, 1988.

Priestley, J.B. *Margin Released: A writer's reminiscences and reflections*, London, 1962.

Pugsley, Chris, 'The Magic of Moving Pictures: Film making 1895–1918', in Diane Pivac, Frank Stark and Lawrence McDonald, eds, *New Zealand Film: An illustrated history*, Wellington, 2011, pp. 29–51.

Racy, Ali Jihad, 'Record Industry and Egyptian Traditional Music: 1904–1932', *Ethnomusicology*, 20, 1, 1976, pp. 23–48.

Radick, Gregory, 'Primate Language and Playback Experiment in 1890 and 1980', *Journal of the History of Biology*, 38, 3, 2005, pp. 461–93.

Radick, Gregory, *The Simian Tongue: The long debate about animal language*, Chicago, 2007.

Rath, Richard Cullen, *How Early America Sounded*, Ithaca, 2003.

Read, Oliver and Walter L. Welch, *From Tin Foil to Stereo: Evolution of the phonograph*, 2nd edn, Indianapolis, 1976.

Reese, Henry, '"The World Wanderings of a Voice": Exhibiting the cylinder phonograph in Australasia', in Joy Damousi and Paula Hamilton, eds, *A Cultural History of Sound, Memory, and the Senses*, pp. 25–39.

Reeves, Nicholas, 'Through the Eye of the Camera: Contemporary cinema audiences and their experience of war in the film *Battle of the Somme*', in Hugh Cecil and Peter Liddle, eds, *Facing Armageddon: The First World War experience*, London, 1996, pp. 780–98.

Reiss, Eric, *The Compleat Talking Machine: A guide to the restoration of antique phonographs*, New York, 1986.

Renwick, William, ed., *Creating a National Spirit: Celebrating New Zealand's Centennial*, Wellington, 2004.

Renwick, William, *Scrim: The man with a mike*, Wellington, 2011.

Rice, Geoffrey W., ed., *The Oxford History of New Zealand*, 2nd edn, Auckland, 1992.

Riley, Matthew, *Musical Listening in the German Enlightenment: Attention, wonder and astonishment*, Aldershot, 2004.

Robb, Brian, *Silent Cinema*, Harpenden, 2010.

Robertson, D.W., 'The Noises of Spectators, or the Spectator as Additive to the Spectacle', in Richard Abel and Rick Altman, eds, *The Sounds of Early Cinema*, Bloomington, 2001, pp. 183–91.

Robertson, Emma, 'I Get a Real Kick Out of Big Ben: BBC versions of Britishness on the Empire and General Overseas Service 1932–1948', *Historical Journal of Film, Radio and Television*, 28, 4, 2008, pp. 459–73.

Robinson, Howard, *A History of the Post Office in New Zealand*, Wellington, 1964.

Robinson, Penny, *A Grand Victorian Lady: The life and times of the Wanganui Municipal Opera House 1899–1999*, Wanganui, 1999.

Roddam, Catherine, 'Enrico Caruso's Landmark Recordings', *History Today*, 52, 4, 2002, pp. 6–7.

Roeder, George H., 'Coming to Our Senses', *The Journal of American History*, 81, 3, 1994, pp. 112–22.

Rogoff, Rosalind, 'Edison's Dream: A brief history of the kinetophone', *Cinema Journal*, 15, 2, 1976, pp. 58–68.

Rondeau, René, *Tinfoil Phonographs: The dawn of recorded sound*, Corte Madera, CA, 2001.

Rondeau, René, 'The Victor Auxetophone', *In the Groove*, 25, 9, 2000, pp. 4–6, 14–15.

Rorke, Jinty, 'A.N. Whitehouse: An early film pioneer', *Historical Review: Bay of Plenty Journal of History*, 32, 1, 1994, pp. 17–23.

Russell, Lorraine, *Family Hold Back: A 1930s childhood*, Wellington, 2001.

Russo, Alexander, 'Defensive Transcriptions: Radio networks, sound-on-disc recording, and the meaning of live broadcasting', *Velvet Light Trap*, 54, 1, 2004, pp. 4–17.

Ruth, Richard Cullen, *How Early America Sounded*, Ithaca, 2003.

Salkeld, Brian, 'The Dancing Decade: 1920–1930', *Stout Centre Review*, 2, 4, 1992, pp. 3–12.

Samples, Mark C., 'The Humbug and the Nightingale: P.T. Barnum, Jenny Lind, and the branding of a star singer for American reception', *Musical Quarterly*, 99, 3/4, 2016, pp. 286–320.

Sandoz, Mari, *The Christmas of the Phonograph Records: A recollection*, Lincoln, NE, 1966.

Sarkar, Tarpan, Robert Mailloux, Arthur O. Oliner, M. Salazzar-Palma and Dipak L. Senqupta, *History of Wireless*, Hoboken, 2006.

Sartre, Jean-Paul, *The Words*, New York, 1964.

Scannell, Paddy, *A Social History of British Broadcasting: Serving the nation 1922–1939*, Oxford, 1991.

Scapalo, Dean, *The Complete New Zealand Music Charts 1966–2006: Singles, albums, DVDs, compilations*, Wellington, 2007.

Schafer, R. Murray, *The Soundscape: Our sonic environment and the tuning of the world*, Rochester, VT, 1994.

Schmidt, Leigh Eric, *Hearing Things: Religion, illusion, and the American enlightenment*, Cambridge, MA, 2000.

Scholes, Percy, *Learning to Listen by Means of the Gramophone*, London, 1921.

Scholes, Percy, *The Listener's History of Music: A book for any concert-goer, gramophonist or radio listener, providing also a course of study for adult classes in the appreciation of music*, London, 1929.

Scholes, Percy, *Music, the Child and the Masterpiece: A comprehensive handbook of aims and methods in all that is usually called 'musical appreciation'*, London, 1935.

Scholes, Percy, *Practical Lesson Plans in Musical Appreciation by Means of the Gramophone*, London, 1933.

Schroeder, Ralph, *Rethinking Science, Technology and Social Change*, Stanford, 2007.
Schwartz, Hillel, *Making Noise: From Babel to the Big Bang & beyond*, New York, 2011.
Sconce, Jeffrey, *Haunted Media: Electronic presence from telegraphy to television*, Durham, NC, 2000.
Scott, Bonnie Kime, 'The Subversive Mechanics of Woolf's Gramophone in *Between the Acts*', in Pamela L. Gaughie, ed., *Virginia Woolf in the Age of Mechanical Production*, New York, 2000, pp. 97–113.
Scott, D. Tracers, 'Sound Studies for Historians of New Media', in David W. Park, Nicholas W. Jankowski and Steve Jones, eds, *The Long History of New Media: Technology, historiography, and contextualising newness*, New York, 2011, pp. 75–88.
Seifert, Marsha, 'Aesthetics, Technology and the Capitalization of Culture: How the talking machine became a musical instrument', *Science in Context*, 8, 2, 1995, pp. 417–50.
Sharma, Bhesham R., *Music and Culture in the Age of Mechanical Reproduction*, New York, 2000.
Sharp, G.N., 'The Gramophone in Musical Education', *Music & Letters*, 19, 2, 1938, pp. 199–202.
Sheirtcliff, Reo, 'Dancing in the Dark: A memoir of Epi Shalfoon', *Music in New Zealand*, 10, 1990, pp. 40–45.
Shepard, Deborah, *Reframing Women: A history of New Zealand film*, Auckland, 2000.
Shepard, Deborah, ed., *Between the Lives: Partners in art*, Auckland, 2005.
Shepard, Deborah, 'Shadow-Play: The film-making partnership of Rudall and Ramai Hayward', in Deborah Shepard, ed., *Between the Lives: Partners in art*, Auckland, 2005, pp. 113–35.
Shepherd, John and Peter Wicke, *Music and Cultural Theory*, Cambridge, 1997.
Shirley, Henry, *Just A Bloody Piano Player*, Auckland, 1971.
Shuker, Roy, *Educating the Workers? A history of the Workers' Education Association in New Zealand*, Palmerston North, 1984.
Shuker, Roy, 'Popular Culture, Moral Panic and Schooling: New Zealand youth at the movies', in John Codd, Richard Harker and Roy Nash, eds, *Political Issues in New Zealand Education*, 2nd edn, Palmerston North, 1990, pp. 235–52.
Siefert, Marsha, 'Aesthetics, Technology, and the Capitalization of Culture: How the talking machine became a musical instrument', *Science In Context*, 8, 2, 1995, pp. 417–49.
Siefert, Marsha, 'The Audience at Home: The early recording industry and the marketing of musical taste', in James D. Ettema and Charles D. Whitney, eds, *Audiencemaking: How the media create the audience*, Thousand Oaks, CA, 1994, pp. 186–214.
Sigley, Simon, 'Film Culture: Its development in New Zealand 1929–1972', PhD thesis, University of Auckland, 2003.
Sigley, Simon, 'How *The Road to Life* (1931) Became the Road to Ruin: The case of the Wellington Film Society in 1933', *New Zealand Journal of History*, 42, 2, 2008, pp. 196–215.

Simpson, Adrienne, *The Greatest Ornaments of their Profession: The New Zealand tours by the Simonsen Opera Companies 1876–1889*, Christchurch, 1993.
Simpson, Adrienne, 'On Tour with the Simonsens', in Adrienne Simpson, ed., *Opera in New Zealand: Aspects of history and performance*, Wellington, 1990, pp. 19–32.
Simpson, Adrienne, ed., *Opera in New Zealand: Aspects of history and performance*, Wellington, 1990.
Simpson, Adrienne, *Opera's Farthest Frontier: A history of professional opera in New Zealand*, Auckland, 1996.
Sinclair, Keith, *A History of New Zealand*, 2nd edn, Auckland, 1988.
Sklar, Robert, *Movie-Made America: A cultural history of American movies*, New York, 1994.
Smilor, Raymond W., 'Cacophony at Thirty-fourth and Sixth: The noise problem in America 1900–1930', *American Studies*, 18, 1977, pp. 23–38.
Smilor, Raymond W., 'Personal Boundaries in the Urban Environment: The legal attack on noise 1865–1930', *Environmental Review*, 3, 3, 1979, pp. 24–36.
Smilor, Raymond W., 'Toward an Environmental Perspective: The anti-noise campaign 1893–1932', in Martin V. Melosi, ed., *Pollution and Reform in American Cities 1879–1930*, Austin, 1980, pp. 135–51.
Smith, Bruce, *The Acoustic World of Early Modern England: Attending to the O-factor*, Chicago, 1999.
Smith, Jacob, 'The Frenzy of the Audible: Pleasure, authenticity and recorded laughter', *Television & New Media*, 6, 1, 2005, pp. 23–47.
Smith, Jacob, *Spoken Word: Postwar American phonograph cultures*, Los Angeles, 2011.
Smith, Mark M., ed., *Hearing History: A reader*, Athens, GA, 2004.
Smith, Mark M., *Listening to Nineteenth Century America*, Chapel Hill, NC, 2001.
Smith, Mark M., 'Making Sense of Social History', *Journal of Social History*, 37, 1, 2003, pp. 165–86.
Smith, Mark M., 'Producing Sense, Consuming Sense, Making Sense: Perils and perspectives for sensory history', *Journal of Social History*, 40, 4, 2007, pp. 841–57.
Smith, Mark M., *Sensing the Past: Seeing, hearing, smelling, tasting, and touching in history*, Berkeley, CA, 2007.
Smith, Merritt Roe and Leo Marx, eds, *Does Technology Drive History?: The dilemma of technological determinism*, Cambridge, MA, 1998.
Smith, Nick, 'The Splinter in Your Ear: Noise as the semblance of critique', *Culture, Theory & Critique*, 46, 1, 2005, pp. 53–59.
Smith, Philippa Mein, *A Concise History of New Zealand*, Cambridge, 2007.
Smith, Stephen, *The Samoa (N.Z.) Expeditionary Force 1914–1915*, Wellington, 1924.
Smithies, James, 'The History of Technology and the History of New Zealand', *Journal of New Zealand Studies*, 4/5, 2005/2006, pp. 111–28.
Soler, Janet, 'Renegotiating Cultural Authority: Imperial culture and the New Zealand primary school curriculum in the 1930s', *History of Education*, 35, 1, 2006, pp. 11–25.

Somerville, Keith, *Radio Propaganda and the Broadcasting of Hatred*, London, 2012.
Sowry, Clive, 'Edison's Kinetoscope', *The Big Picture*, 8, 1996, pp. 11–12.
Sowry, Clive, 'Exhibiting Films in Jubilee Year', *The Big Picture*, 10, 1997, pp. 20–21.
Sowry, Clive, 'The Illusionist and the Cinematographe', *The Big Picture*, 11, 1996, pp. 20–21.
Sowry, Clive, 'The Kinematograph arrives in New Zealand', *The Big Picture*, 9, 1996, pp. 22–23.
Sowry, Clive, 'Macmahon's Cinematograph', *The Big Picture*, 10, 1996, pp. 22–23.
Sowry, Clive, 'Non-Fiction Films: Between the Wars', in Diane Pivac, Frank Stark and Lawrence McDonald, eds, *New Zealand Film: An illustrated history*, Wellington, 2011, pp. 79–101.
Sowry, Clive, 'Northcote's Kinematograph on Tour: Letters from an early film exhibitor', *The Big Picture*, 13, 1997, pp. 2–25.
Speedy, Squire L., *The Picturedrome Fun Merchant: Anecdotes of the life and times of the Picturedrome cinema and dancehall at Milford during the era of L.L. Speedy, 1922-1938*, Auckland, 1992.
Spehr, Paul, *The Man Who Made the Movies: W.K.L. Dickson*, New Barnet, Herts, 2008.
Spehr, Paul, 'Movies and the Kinetoscope 1890–1895', in André Gaudreault, ed., *American Cinema 1880-1909: Themes and variations*, New Brunswick, NJ, 2009, pp. 22–44.
Spittle, Gordon, 'Johnny Cooper: The original Maori Cowboy', *Music in New Zealand*, 24, 1994, pp. 43–45.
Spittle, Gordon, *Counting the Beat: A history of New Zealand song*, Wellington, 1997.
Spittle, Gordon, *The Tex Morton songbook*, Auckland, 2008.
Spring, Katherine, 'Pop Go the Warner Bros. et al.: Marketing film songs during the coming of sound', *Cinema Journal*, 48, 1, 2008, pp. 68–89.
Spring, Katherine, 'To Sustain Illusion is All that is Necessary: The authenticity of song performance in early American sound cinema', *Film History*, 23, 3, 2011, pp. 285–99.
Staff, Bryan and Sheran Ashley, *For the Record: A history of the recording industry in New Zealand*, Auckland, 2002.
Stafford, Jane and Mark Williams, *Maoriland: New Zealand literature 1872-1914*, Wellington, 2006.
Staiger, Janet and Sabine Hake, eds, *Convergence Media History*, New York, 2009.
Standish, Russell, *Eltham: One hundred years*, Eltham, 1984.
Steffen, David J., *From Edison to Marconi: The first thirty years of recorded music*, Jefferson, NC, 2005.
Sterne, Jonathan, *The Audible Past: Cultural origins of sound reproduction*, Durham, NC, 2003.
Sterne, Jonathan, 'Being "In the True" of Sound Studies', *Sound, and the Moving Image*, 2, 2, 2008, pp. 163–67.
Sterne, Jonathan, 'A Machine to Hear Them: On the very possibility of sound's reproduction', *Cultural Studies*, 15, 2, 2001, pp. 259–94.
Sterne, Jonathan, *The Sound Studies Reader*, Oxford, 2012.

Sterne, Jonathan, 'Sounds like the Mall of America: Programmed music and the architectonics of commercial space', *Ethnomusicology*, 41, 1, 1997, pp. 22–50.
Stokes, John W., *The Golden Age of Radio in the Home*, Invercargill, 1986, p. 28.
Stranger, Ralph, *Wireless: The modern magic carpet*, 5th edn, London, 1933.
Sturm, Terry, ed., *The Oxford History of New Zealand Literature in English*, 2nd edn, Oxford, 1998.
Suárez, Juan A., 'T.S. Eliot's *The Waste Land*, the Gramophone and the Modernist Discourse Network', *New Literary History*, 32, 2001, pp. 747–68.
Suisman, David, *Selling Sounds: The commercial revolution in American music*, Cambridge, MA, 2009.
Suisman, David, 'Sound, Knowledge, and the "Immanence of Human Failure": Rethinking musical mechanization through the phonograph, the player-piano, and the piano', *Social Text*, 28, 1, 2010, pp. 13–34.
Suisman, David and Susan Strasser, eds, *Sound in the Age of Mechanical Reproduction*, Philadelphia, 2010.
Sullivan, Jim, *An Independent Air: The story of the Otago Radio Association and New Zealand's oldest station 4XD Dunedin*, Dunedin, 1997.
Sullivan Jim, *As I Remember: Stories from 'Sounds Historical'*, Auckland, 1999.
Sullivan Jim, *As I Remember Vol. 2: Stories from 'Sounds Historical'*, Auckland, 2000.
Sullivan, Jim, *Canterbury Voices*, Christchurch, 2007.
Symes, Colin, *Setting the Record Straight: A material history of classical recording*, Middletown, CT, 2004.
Symes, Colin, 'From *Tomorrow's Eve* to High Fidelity: Novel responses to the gramophone in twentieth century literature', *Popular Music*, 24, 2, 2005, pp. 193–206.
Symes, Colin, 'A Sound Education: The gramophone and the classroom in the United Kingdom and the United States 1920–1940', *British Journal of Music Education*, 21, 2, 2004, pp. 163–78.
Tairawhiti Maori Association, *Echoes of the Pa: Tairawhiti Maori Association's research proceedings 1931–2*, Gisborne, 1932.
Taruskin, Richard et al., 'The Limits of Authenticity: A discussion', *Early Music*, 12, 1984, pp. 3–25, 523–25.
Taussig, Michael, *Mimesis and Alterity: A particular history of the senses*, New York, 1993.
Tayler, Douglas, *A Complete Scheme of School Music Related to Human Life* Wellington, 1927.
Tayler, Douglas, 'A Musical Query', *Education Gazette*, 1 May 1928, p. 67.
Taylor, Linda L., 'Commerce and Poetry Hand in Hand: Music in American department stores 1880–1930', *Journal of the American Musicological Society*, 45, 1, 1992, pp. 75–120.
Taylor, Timothy D., *Strange Sounds: Music, technology and culture*, New York, 2001.
Thompson, Emily, 'Machines, Music and the Quest for Fidelity: Marketing the Edison phonograph in America 1877–1925', *The Musical Quarterly*, 79, 1, 1995, pp. 131–71.
Thompson, Emily, *The Soundscape of Modernity: Architectural acoustics and the culture of listening in America*, Cambridge, MA, 2002.

Thompson, Kristin and David Bordwell, *Film History: An introduction*, New York, 1994.

Thomson, Alistair G., ed., *Phonographs & Gramophones: A commemorative catalogue of the exhibition held at the Royal Scottish Museum from 2nd July–2nd October 1977 to celebrate the centenary of Thomas Edison's invention of the phonograph*, Edinburgh, 1977.

Thomson, Alistair G., ed., *Phonographs & Gramophones: A symposium organised by the Royal Scottish Museum in connection with the exhibition Phonographs and Gramophones and the Centenary of the Invention of the Phonograph by Thomas Alva Edison*, Edinburgh, 1977.

Thomson, John Mansfield, *A Distant Music: The life and times of Alfred Hill 1870–1960*, Auckland, 1980.

Thomson, John Mansfield, *The Oxford History of New Zealand Music*, Auckland, 1991.

Thomson, John Mansfield, 'Sight and Sound: Exhibitions & New Zealand music 1865–1940', *Music in New Zealand*, 16, 1992, pp. 34–60.

Thomson, John Mansfield, 'A Triumph for Instrumental Music of the Highest Type: From the orchestra to the Besses O' the Barn Band', in J.M. Thomson, ed., *Farewell Colonialism: The New Zealand Exhibition Christchurch, 1906–07*, Palmerston North, 1998, pp. 79–93.

Thomson, John Mansfield, ed., *Farewell Colonialism: The New Zealand Exhibition Christchurch, 1906–07*, Palmerston North, 1998.

Thorburn, David and Henry Jenkins, eds, *Rethinking Media Change: The aesthetics of transition*, Cambridge, MA, 2003.

Thwaites, Ian, 'Arthur Collyns: An Auckland broadcasting pioneer', *Auckland-Waikato Historical Journal*, 80, 2002, pp. 1–17.

Toogood, Selwyn, *Out of the Bag*, Auckland, 1979.

Toop, David, *Haunted Weather: Music, silence and memory*, London, 2004.

Toop, David, *Sinister Resonance: The mediumship of the listener*, New York, 2010.

Toulmin, Stephen, *Cosmopolis: The hidden agenda of modernity*, Chicago, 1990.

Trigg, David, *Clockwork Music Hall*, Ilfracombe, 1976.

Turley, Alan, 'The Incredible Tex Morton', *New Zealand Memories*, 37, 2002, pp. 4–12.

Utting, Alec, ed., *Life, Laughter and Love in the Early Years: Remembering earlier times in Birkenhead and Birkdale*, Auckland, 2004.

Vallee, Mickey, 'Flat Disc Phonographs and the Injunction of the Second Master', *Journal of Historical Sociology*, 24, 3, 2011, pp. 245–65.

Vancour, Shawn, 'Arnheim on the Radio: *Materialtheorie* and beyond', in Scott Higgins, ed., *Arnheim for Film and Media Studies*, New York, 2011, pp. 177–94.

Vancour, Shawn, 'Popularizing the Classics: Radio's role in the American music appreciation 1922–34', *Media, Culture & Society*, 31, 2, 2009, pp. 189–307.

Voegelin, Salome, *Listening to Noise and Silence: Towards a philosophy of sound art*, New York, 2010.

Walker, David, *Stratford: Shakespearean town under the mountain, a history*, Wellington, 2005.

Walker, Jesse, *Rebels on the Air: An alternative history of radio in America*, New York, 2001.

Waller, Gregory, 'Robert Southard and the History of Traveling Film Exhibition', *Film Quarterly*, 57, 2, 2003, pp. 2–14.
Watt, John, ed., *Radio Variety*, London, 1939.
Weber, Heike, 'Head Cocoons: A sensori-social history of earphone use in West Germany 1950–2010', *Senses & Society*, 5, 3, 2010, pp. 339–63.
Weber, William, 'Did People Listen in the Eighteenth Century?', *Early Music*, 25, 4, 1997, pp. 678–92.
Weheliye, Alexander G., *Phonographies: Grooves in sonic Afro-modernity*, Durham, NC, 2005.
White, Shane and Graham White, 'At Intervals I was Nearly Stunned by the Noise He Made: Listening to African American religious sound in the era of slavery', *American Nineteenth Century History*, 1, 1, 2000, pp. 34–61.
Wierzbicki, James, *Film Music: A history*, New York, 2009.
Willliams, Alan, 'Historical and Theoretical Issues in the Coming of Recorded Sound to the Cinema', in Rick Altman, ed., *Sound Theory, Sound Practice*, New York, 1992, pp. 126–37.
Willis, Margaret, *Aunt Gwen of 2YA*, Auckland, 2008.
Wilson, A.C., *Wire & Wireless: A history of telecommunications in New Zealand 1890–1987*, Palmerston North, 1994.
Wilson, Eric, 'Plagues, Fairs, and Street Cries: Sounding out society and space in early modern London', *Modern Language Studies*, 25, 2, 1995, pp. 1–42.
Wilson, H.L., *Music and the Gramophone and Some Master Recordings*, London, 1926.
Wilson, Sarah, 'Gertrude Stein and Radio', *Modernism/Modernity*, 11, 2, 2004, pp. 261–78.
Wimbush, Roger, ed., *The Gramophone Jubilee Book 1923–1973*, Harrow, 1973.
Wipplinger, Jonathan, 'The Aural Shock of Modernity: Weimar's experience of jazz', *The Germanic Review*, 82, 4, 2007, pp. 299–320.
Wojcik, Pamela Robertson and Arthur Knight, eds, *Soundtrack Available: Essays on film and popular music*, Durham, NC, 2001.
Woledge, C.E., 'Talking Machine Memories', *The Phonographic Record*, 2, 2, 1966, pp. 15–16.
Wondrich, David, *Stomp and Swerve: American music gets hot 1843–1924*, Chicago, 2003.
Woolf, D.R., 'Speech, Text and Time: The sense of hearing and the sense of the past in Renaissance England', *Albion*, 18, 2, 1986, pp. 159–93.
Wurtzler, Steve J., *Electric Sounds: Technological change and the rise of corporate mass media*, New York, 2007.
Yablon, Nick, 'Echoes of the City: Spacing sound, sounding space, 1888–1916', *American Literary History*, 19, 3, 2007, pp. 629–60.
Yagou, Artemis, 'Shaping Technology for Everyday Use: The case of the radio set design', *The Design Journal*, 5, 1, 2002, pp. 2–13.
Zielinski, Siegfried, *Deep Time of the Media: Toward an archaeology of technical hearing and seeing*, Cambridge, MA, 2006.

B. WEBSITES

Alexandratos, Michael, 'Rotorua Maori Choir: Profile', *Audioculture*, www.audioculture.co.nz/people/rotorua-maori-choir

Archer, John, 'Ten Guitars', *New Zealand Folk Song*, http://folksong.org.nz/tenguitars/index.htm

Archer, John, 'He Puru Taitama', *New Zealand Folk Song*, www.folksong.org/he puru/index.html

Barr, Mary and Jim Barr, 'The Kid From Timaru', *The New Zealand Film Archive*, www.filmarchive.org.nz/tracking-shots/close-ups/KidFromTimaru.html

Beaglehole, John, 'John Beaglehole to parents, 28 June 1929', John Cawte Beaglehole Letters, *New Zealand Electronic Text Centre*; www.nzetc.victoria.ac.nz/tm/scholarly/tei-JCB-081.html

Christchurch City Libraries, *Ernest John Bell: 1885?–1971*, http://christchurchcitylibraries.com/Heritage/People/B/BellErnestJ/; *Remembering Jack and Edna*, online, http://library150.com/Articles/ErnestBell/

Dougherty, Ian, 'Bell, Francis Wirgman Dillon 1896–1987; Bell, Margaret Brenda 1891–1979'. Dictionary of New Zealand Biography, www.teara.govt.nz/en/biographies/4b20/1

Downes, Peter, 'Dech, Gil 1897–1974', Dictionary of New Zealand Biography, www.teara.govt.nz/en/biographies/5d13/1

Downes, Peter, 'Drummond, David Archibald Victor Clive 1890–1978', Dictionary of New Zealand Biography, www.teara.govt.nz/en/biographies/4d20

First Sounds, 'World's Oldest Recording Made Available Online', *First Sounds*, www.firstsounds.org

Hawkey, Rachel, 'Griffiths, Thomas Vernon 1894–1985', Dictionary of New Zealand Biography, www.teara.govt.nz/en/biographies/4g21/1

Malcolm, Joe, 'Hato, Ana Matawharua 1907–1953', Dictionary of New Zealand Biography, www.teara.govt.nz/en/biographies/4h19/1

Matthews, Mary, 'Our Radio', *Sounds Historical*, www.radionz.co.nz/national/programmes/soundshistorical/20080302

Pivac, Doane, 'Coubray, Edwin: Biography', Dictionary of New Zealand Biography, www.TeAra.govt.nz/en/biographies/5c38/1

Powell, Jonathan, 'Skryabin, Aleksandr Nikolayevich', *Grove Music Online: Oxford Music Online*, www.oxfordmusiconline.com/subscriber/article/grove/music/25946

Shalfoon, Reo, 'Shalfoon, Gareeb Stephen: Biography', Dictionary of New Zealand Biography, www.teara.govt.nz/en/biographies/4s22/1

Skelton, L.R., 'Hayward, Rudall Charles Victor: Biography', Dictionary of New Zealand Biography, www.teara.govt.nz/en/biographies/4h22/1

Sousa, John Philip, 'The Menace of Mechanical Music', *Appleton's Magazine*, 8, 1906, pp. 278–84, cited in Patrick Feaster, 'Phonozoic Text Archive, Document 155', *Phonozoic*, www.phonozoic.net/n0155.htm

Sowry, Clive, 'Whitehouse, Alfred Henry 1856–1929', Dictionary of New Zealand Biography, www.teara.govt.nz/en/biographies/2w16/1

Spittle, Gordon, 'Morton, Tex 1916–1983', *Dictionary of New Zealand Biography*, www.TeAra.govt.nz/en/biographies/5m59/1

Traue, J.E., 'Buick, Thomas Lindsay 1865–1938', *Dictionary of New Zealand Biography*, www.teara.govt.nz/en/biographies/3b57/1

Walsh, G.P., 'Narelle, Marie (Molly) (1870–1941)', *Australian Dictionary of Biography*, http://adb.anu.edu.au/biography/narelle-marie-molly-13126/text23753

Ward, Aleisha, 'New Zealand's First Jazz Recording', *Audioculture*, www.audioculture.co.nz/scenes/new-zealand-s-first-jazz-recording

Watson, Chris, '*Frances of Feilding* (Lee Hill, 1928) – A Community Comedy: New Zealand's populist answer to Hollywood', *Screening the Past*, 8, www.latrobe.edu.au/screeningthepast/firstrelease/fr1199/cwfr8c.htm

C. AUDIO RECORDINGS

Hato, Ana and Deane Waretini, *Ana Hato Raua ko Deane Waretini: Legendary recordings 1927–1949*, compact disc, Wellington, 1995.

The Tahiwis, *The Tahiwis: Historic 1930 recordings by Te Whanau Tahiwi*, compact disc, Wellington, 1998.

INDEX

Page numbers in **bold** refer to illustrations.

acetate discs 75
Admiralty, London 104
Adorno, Theodor 16, 17, 74, 75
advertising: advertising power of talkies 184, 185; in digital audio services 8; on radio 121; to sell gramophones, radios and recordings 21, 22, 35, 36, 37, 38, 39, 41, 44, 46, 47, 51, 71, 144
aerials, radio 91, **92**, 94, 116
aesthetics, musical 14
Alhambra Theatre, Dunedin 172–73
Allardyce, Allan 144
aluminium discs 75–76
Amalgamated Wireless Company of Australasia Limited 107
Amberol cylinders 46
America *see* United States
Anderson, Barbara 124
Andrews, George 134
Anthony, Frank S., *Mark and Gus* 147
anthropologists, preservation of music by recording 14–15
Anzac Wireless Squadron 110–11
Archibald, Douglas **26**, 26–28, **29**, 31, 32, 166
Armistice Day ceremony broadcast 150
Arnhein, Rudolf 122, 123, 148, 195
associative listening 122
Athenaeum, Dan 38
Atkinson, Karl 73
Auckland 107, 108; Britannia Theatre 176; Civic Theatre 190–91; dancing 68; Empire Theatre 175, **176**; phonograph exhibitions 27–28; radio station 91; silent films 166, 167, 175–76, **176**, 179; St James Theatre 191; The Talkeries 35
Auckland Exhibition, 1898 166
Auckland Film Society (AFS) 195
Auckland Star 144, 195; 'A Football Plan' 145, **146**; 'A Grandstand Seat by Radio' 144–45, **146**; 'The Magic Spark' 116
audio technologies 7–8, 9, 11–12, 18, 201–02, 204; concerns about 202; *see also* digital audio technologies; and individual technologies, e.g. radio; recordings; silent films
audiophiles 46
Aunt Betty, 1YA 127

Aunt Daisy (Maud Basham, née Taylor) 200
Aunt Edna, 3YA (Edna Pearce) 142
Aunt Gwen, 2YA (Gwen Shepherd) 119–20, **120**, 142, 143–44
Australia 37, 78, 83, 150, 177; phonographs 22, 25, 26, 40; radio programmes 128, 142, 147, 148; radio stations 96, 97, 134; silent films 159, 160, 163; singers touring New Zealand 37, 56; World War I troops 104, 110, 115
Australian Commonwealth Naval Board 104
Auxetophone 38
Awanui radio station 91
Awarua radio station 91
Ayers, Reynold 181

band music 137
Bartlett's photographic studio 155
Bartok, Bela 60
Basham, Frederick 200
Basham, Maud (née Taylor) ('Aunt Daisy') 200
Battershill, Eric (pseud. Eric Dare) 87, 88, **88**, 89, 91, 92, 115
Baxter, Archibald 51–52
Baxter, Harry 158, 159
BBC 126, 129, 150, 151; 'Children's Hour' 142; Empire Service 150, 151
Beaglehole, J.C. 195
Belich, James 10, 193
Bell, Alfred 92
Bell, Brenda 92
Bell, Ernest (Uncle Jack, 3YA) 142–43
Bell, Frank 92
Bellingham, J. 131–32
Berliner, Emil 26
Besant, Annie 102
Bierre, Epi 184, 185
Billings, Elsie 72
Biograph Company 167
Bio-Tableau film company 171–72
Birkenhead 179
Blenheim, Lyceum Hall 21
blues 66
Bohanna, Peter J. 34–35, 201
Boothman, Alfred 160
Boyd, Olive 50

278

Brandon, Cyril 93–94
Brescians 174–75
Britain: British-based culture of New Zealand, as opposed to American 193–94, 195–97; first two-way radio conversation with New Zealand 92; high culture 18; New Zealand radio broadcasting of events 150–51; phonograph exhibitions 22, 26
Britannia Theatre, Auckland 176
British Biograph Company 159, 162, 167, **168**
British Dominion Films Ltd 161
British Music Society (BMS) Gramophone Groups 60, 62–64, 67, 68, 71
British Photographic Studio 167
Broadcasting Amendment Bill 1935 134
Broadway Melody (film) 190, 192
Brunswick Panatrope 64
Brunswick Records 191
Buckman, Rosina 56
Buick, Lindsay, *The Romance of the Gramophone* 67, 75
Bull, Stanley 127
Bullock, H.N. 137
Burke, E.J. 177

cabarets 57, 75
Calloway, Cab 185
Campbell, Charlie 60
Campbell, Ewan 63
'canned music' 22, 39, 74, 125, 129–30, 165, 171, 186
Canterbury (University) College 60, 62, 90
Carnegie Corporation 59, 64
Caruso, Enrico 14, 16, 35, 38
CDs 7, 8
celebrity culture 151
censorship 197
'Chalita' 68, **69**, 71, **71**
HMS *Challenger* 95
Charles Begg & Sons 34, 35, 39, 43, 63; gramophone and phonograph department 38
Chatham Islands, radio station 91–92, **92**, 104
children: birthday greeting requests on radio 142–43; fears about harmful effects of radio 141–42; influence of films 141; radio programmes 142–44
Christchurch 108; Grand Theatre 179; phonograph exhibitions 21, 24; The Talkeries 35
Christchurch Exhibition, 1906–07 99
Christchurch Kinematograph Syndicate 158
Christian, Marion 127
Chronicles of the NZEF 50–51

Chronophone 168–69, 181
Chubb, Rowland 167, **168**
cinemas: attendances 186; impacts of the talkies 198–99; new spaces for hearing music 13–14, 18; number of cinemas 189; over-supply of venues 189; projection rooms 190; purpose-built 169–70, 175–76, **176**, 183, 190–91; sound systems 189–91, 198–99; use of theatres and halls before dedicated cinemas 174–75, 190; *see also* films in New Zealand
Cinematograph Films Act 1928 197
Cinephone 169
Civic Theatre, Auckland 190–91
Clark, George 127
Clarke, John, 'Fred Dagg' 147
Clarke, Marcus 160
classical music 13, 16, 17, 18, 35, 36, 48, 56–57, 58, 59–60, 62, 63–64, 65, 130, 131, 132, 139, 141; complete sets of works 67; in silent films 176–77
Clauson, Gerard 112
Collins, Ken 96, 130
Columbia Records **61**, 68, **69**, **70**, **71**, 78, 79, **80**, **81**, 83, **192**
commodification of music 17
community films 187–88, **188**
Cooper, Johnny ('The Maori Cowboy') 66
Corliss, Valerie 62
Cornford, H.A. 87
Coubray, Edwin 186
country music 65–66, 75–76
Courtis, Burall 111
Coyne, Margaret 58
crooning 72, 74
Crosby, Bob, 'Woman on My Weary Mind' 72
Croucher, Richard 111
cultural nationalism, New Zealand 193–98
culture *see* global consumer and leisure cultures; high culture; low culture; popular culture

'Dad and Dave from Snake Gully' 147, **149**
Daley, Caroline 58
dance music 68, **69**, **70**, **71**, 71–72, 141, 184
Dawson, Peter 56
Davies, Joseph 120
Day, Patrick 147
De Groen's Vice-regal Orchestra 175
Decca 12, **13**
Dech, Gil 78
Defence Department 108–09
Dennes Bros 35
Depression 62, 121

279

Diamond Discs (Edison Company) 65
digital audio technologies 7, 8–9, 11, 18, 201, 202, 203
dimensional listening 122
Dominion College of Radio-telegraphy Ltd 107–08, **108**, 109
Dorsey, Jimmy 185
Douglas, Susan 122
Down on the Farm (film) 187
Downes, Peter 78
Dresden Piano Company 35
drilling and marching 58
Drummond, Clive 96, 104, 115, 118, 119, 120
Dunedin 169, 187; Alhambra Theatre 172–73; Charles Begg & Sons 38; phonograph exhibitions 26; The Talkeries 35; wireless telegraphy 93–94
Dunedin Athenaeum 94, 127
Dunedin Gramophone Group 63–64
durability of sound 12, 14–16, 18, 24, 204
Durium record label 15, **15**

earphones 95–96, **97**, 203
Edison Company **37**, 46, 65
Edison Phonograph **47**
Edison, Thomas 9, 22, **23**, 24, 26, 28, 31, 32, 54, 155, 156
Edmonds baking powder 49–50, **51**
education: Johnson, William, *The Gramophone in Education* 54, 59; learning dance steps 68, 71–72; music in conjunction with literature, history and geography studies 56; musical education 14, 48, 53–57, **55**, **57**, 59, 73; recordings, learning opportunities for musicians 14, 65–66; school dancing, exercise and drilling 57–58; university music education 59–60; writing and speech 58–59
Education Gazette 55, 58–59
Edwards, C.A. 25
Eltham 190
Empire Theatre, Auckland 175, **176**
entertainment industry 18, 22, 27, 33, 199, 204
entertainment programming on radio 127, 129–30, 131–32, 136, 137, 139
ethnomusicologists, preservation of music by recording 14–15
Europe 40; high culture 18; industrial production of records and films 16, 33
Evening Post 78, 139, 141, 195–96
exhibitions and demonstrations of phonographs 21, 22, 24–28, **25**, **29**, 31, 32, 33, 43

Falconer, Dick 143–44
film announcers and narrators 158–64, 173–74, 175
films: digital technologies 8; European opinions about synchronised sound 195; permanence 16; repetition of sound 16, 204; separation of sound from sources 8, 9, 12
films in New Zealand: American films and film stars 186, 187, 191, 193–98; British films 193–94, 196, 197, 198; censorship 197; film shorts 184–85; first sound films 156, 186–89; influence on speech 59, 195–96, 197–98; learning opportunities for musicians 65, 66; local people and places on screen 187–89; music shorts 184–85, 199; musicians made redundant by talkies 68, 130, 185, 199; New Zealand-made films 184–85, 186–89, 198; news reels 187, 198; objections to 75, 192–97, 198; orchestras after arrival of talkies 186; songs 16, 67, 68, 78, 158, 191–92, 193, 202; standardisation of cinematic sound 186; synchronised sound 168–69, 175, 183, 185–86, 189–90, 192; visual guides to performance techniques 14; *see also* silent films
Findlay, Mary 72
Fisk Jubilee Singers 167
Fitzgerald, A.N. 58–59
For the Term of His Natural Life (film) 160
Forester's Hall, Birkenhead 167
Forester's Hall, Karangahape Road 167
Foxton 182
France, World War I 113–14
Fraser-Jones, Myrtle 74
'Fred and Maggie Everybody' 128, **128**, 129
Fuller's theatre circuit 177

Gaisberg, Fred 83
Gallipoli 161–62, 163
Gaumont Chronophone 168–69
gender roles: broadcast sports 147; high-end audio equipment 46; moving picture technology 158; phonographs 44–45, **45**, 47, 50; piano playing 44; uncertainties 74; wireless telegraphy 92, 93
George & Kersley, Wellington 38
George VI, coronation 150–51
Gibbons, Peter 9–10, 22, 152
Gladstone, William 27, 28
global consumer and leisure cultures 9–10, 58, 69, 100, 157, 203

Glover, Denis 60
'God Defend New Zealand' 56
'God Save the King' 174, 175
Gold Diggers of Broadway (film) 11, 191, 192, 193, 195
Goosen, Albert 56
Grainger, Percy 15
Gramophone 41
Gramophone Company 34, 35, 83
Gramophone Grand machine 39
Gramophone Groups, British Music Society (BMS) 60, 62–64, 67, 68, 71
gramophones: annoyance and irritation 72; in cinemas 13, 165, 169–70; criticisms 74–75; disc-based gramophone 26; educational use 53–66, **57**, 68, 71–72; jokes, stories and anecdotes 48, 50; listening to 12–13, 18, 22; portable 8, 11, **13**, 53, 72; prices 43; in private households 40–41, 43; recitals of recorded music **64**, 64–65, 127; revolution in listening 47–48; separation of sound from sources 8, 9, 52; unexpected uses 22; wartime use 50–52, **51**, **52**; *see also* phonographs
Grand Theatre, Christchurch 179
Greenwood, H. 127
Grey, George, phonograph recording **27**, 27–28
Greymouth 182
Griffen, Mark 182
Griffiths, C.J.W. 21, 22
Griffiths Duo 127
Griffiths, Vernon 62–63
Gronow, Pekka 76

Hagerty, James 161
Ham radio hobbyists ('DXers') 88, 92
Hamilton Talks (film) 187, 188
Hanley, P. 179
Happy St. Georges Company 158
Hare Hongi (Henry H. Stowell) 148
Hargreaves, Eva 190
Harrall, Ted 57–58
Harris, Ambrose 132
Harris, A.R. 129
Harrison, John 38
Hastings Gramophone Group 63
Hathaway, Alfred 95
Hato, Ana 76, **77**, 78, 82, 83
Hawkey, Rachel 63
Hayward, Henry 174, 175
Hayward, Hilda 187
Hayward, Rudall 187–89, **188**
Hayward's Picture Enterprises 175

'He Puru Taitama' 184
hearing 11; *see also* listening; sounds
'hearing the world' 9–10
Hicks, Stanton 93–94
high culture 18, 35, 48, 54–55, 57, 59–60, 63, 64, 74–75, 131, 132, 139, 141, 194–95
Hill, Alfred 177
Hill, John and Laurie 147
'Hine e Hine' 56
Hinemoa (film) 177, 179
HMV (His Master's Voice) 74, **193**; opera recordings 54
Holton, Thomas 35
homes: phonographs and gramophones 40–41, **42**, 43, 44–45, 46–48, **47**, 53, 54, 61, 63, 67, 68; pianos 40–41, 43, 44, 45; and portability of audio technologies 14; radio 12, 14, 53, 123–26, **124**, **125**, 127, 128–29, 135, 139, 141; ways of listening 13, 44–45, **45**
Hongi, Hare (Henry H. Stowell) 148
humour about phonographs and gramophones 48–50, **49**
Humperdinck, Engelbert 177, 179
Hutter, Gordon 145
Hyams 35
Hyde, Robin 174–75

imagination in radio listening 120, 122, 123, 144, 148, 150, 151, 152
Impossible Voyage, An (Voyage à travers impossible) (film) 174–75
Indian–Anglo army, Mesopotamian Campaign 111, 112
informational listening 122
internet 7, 8
Invercargill 38, 175
Iona Presbyterian College for Girls 55, **55**
iPods 8–9, 203
iTunes 7, 8

Jack, Robert 116
Jacobson, N.R. 127
jazz 57, 65, 66, 67, 68, 171, 185; objections to 72–74, 75, 131, 132, 134
Johnson, Eldridge 35
Johnson, William, *The Gramophone in Education* 54, 59, 60
Jones, Ada 36, 37
Jones, Innes 119
Joyce, James 32

Katz, Mark 66
Kāwhia 182
Kelly Gang, The (film) 160

Kerepehi School, Hauraki Plains 58
Kid from Timaru, The (film) 161, **163**
'Kid from Timaru, The' (poem) 161–62
kinematograph films 158
kinetophone 155, **156**, 165, 168
kinetoscope 155, 165, 166
King, Michael 10
Kiwis concert party 83
Kohler, J. 25

Lambert, Constant 17, 74
Lambeth Walk **17**
Langtry, Lily 161
Lanigan, Ted 179
Lauder, Harry 36, 38, 169, 181
Lawson, Douglas 169–70
Lee, John A. 175
leisure culture 9, 10, 58, 69, 100, 157, 203; 'mass mechanical amusement' 74; radio 128–29
Lemke, Claudia 82
Lilburn, Douglas 59, 60, 65
Lind, Jenny 14
listening: classical music 12–13, 55, 64, 65, 68, 71, 74–75; earphones 95–96, **97**; at home 44–45, **45**, 46, **47**, 67; impact of mass production and repetition of sound 16, 17–18, 52; impact of the capability to preserve sounds 14; as means of radicalising and liberating conservative norms 75; modes 12–14, 44–45, 53, 55, 64, 65, 68, 74–75, 204; places and times 117–18, 121–22; popular music 68, 75; private experience 205; revolution brought about by phonographs and gramophones 21–22, 24, 27–28, 32–33, 43, 44–48; school music 55, 56, 57; shaped by social, cultural and personal factors 7; wartime wireless transmissions 104, 105–06, 112, 113; as a way of engaging with the world 18, 203, 204–05; *see also* hearing; radio listening; sounds
listening groups 60–65
Living London (film) 173
Lloyd, Harold 198
Lockington, G.E. 47
Lodge, Oliver, *Signalling Through Space Without Wires* 99
London 76, 173
low culture 18, 65–66
Lucas, Nick ('The Crooning Troubadour') 11, 12, 191

MacMahon, Joseph F. 158, 166
MacMahon, William 27, **29**
Macnamara's Family Band 66
Majestic Theatre Orchestra 127
Major Perry's Biorama Company 180–81
Manning, Nena 158–59
Mansfield, Katherine 47–48
Māori: Hayward's films 189; and Pākehā conception of nationhood 78, 82–83, 150; paternalistic attitudes of Pākehā towards culture 148, 150; radio broadcasts on pronunciation of language 148
'Maori Battalion Marching Song' 201
Māori music: Alfred Hill's use of motifs and music in his work 177, 179; commercial recordings 76, **77**, 78, **79**, **80**, **81**, 82–84, 148, 150, 184; radio broadcasting 148, 150; recordings 15, 56, 76, 82; traditional 76, 82–83; visual marketing of commercial recordings 82
Marbeck, Roger 7
Marbecks music shop, Auckland 7
marching 58
Marconi Company 97, 98, 99, 100
Margery, J. 158
Marschel, Barrie 161–62, **163**
masculinity 93, 110, 147
mass production and replication: American films 186, 193–94; images in silent films 157; music 16, 17–18, 52, 74, 75
Massey, William 49, **49**, 56
Masterton: Gramophone Group 63; The Talkeries 35–36, **36**, 37, **47**
Matheson, Eileen 170
Matthews, Mary 143, 144
Maude, Frederick 112
Mayo, H.R. 95
McDonald, George 108
McEwen, Jack 65
McKenzie, Compton 41
McKinlay, Ernest 81, 83
McKinnon, C.J. 64
McMillan, Llewellyn 110
mechanical instruments 14
Melba, Nellie 35, 38
Méliès, Georges 174–75
Mignon Hornless Gramophone 41
Milford, Picturedrome cinema 170, 179, 190
military radio: German wireless transmissions and stations 103, 104, 105–07, **107**, 111, 112; listeners 104, 105–06, 112, 113; New Zealand wireless operators 105, 107–09, 110, 112–13, **113**; New Zealand Wireless Troop 109–13, **113**; overseas opportunities for learning about new

technologies 114–15; Pacific wireless network 105; Russian messages, Mesopotamian Campaign 112; Turkish radio stations, Mesopotamia 112
Millar, John **49**
Miracle, The (film) 177
Mirams, Gordon 197
Mirror, The 73
mobile phones 8, 201
modernity 22, 74, 100, 184, 185, 188; audio experiences 8, 74, 75, 157, 177, 186, 190–91, 202, 203, 204; characteristics 203–04; concerns about 72, 74; global consumer and leisure cultures 10, 69, 100, 157; plasticity in social organisation, formation and movement of sound 204–05; radio 89, 91, 97, 121
Monckton, C.C.F., *Radio-Telegraphy* 99
Montague, J.F. 148
Montgomery's Pictures and Entertainers 179
More, William (W.D.) 127, 143
Morse code 88, 93, 96, 104, 110, 112, 113
Morton, Tex 65–66, 75–76
Mother Goose (popular show) 181
Mountjoy, W.J. 59
Mules, Betty 134, 148
Murray, Bill 36
music appreciation 48, 53–57, 60–65, 73
Music in New Zealand (MNZ) 62, 63, 64
music retail trade, New Zealand 33, 34–38; alliance with new technologies 39–40; gramophone recitals **64**, 64–65; recitals of recorded sound 37–38; record making 75; slump in sales of musical instruments 39; *see also* names of individual retailers
music shorts 184–85, 199
music teachers 74
musicians: made redundant by talkies 68, 130; remuneration for radio performances 130–31
My Lady of the Cave (film) 189

Napier 169–70; Gaiety Theatre 189
Narelle, Marie 37
National Council of Women 197
national identity, New Zealand 9, 10
National Union of Students, 1936 ball 198
Neilson & Sons 166
Nelson **57**, 63
New Phonogram 22, 40, 50
New Plymouth 35, 181
New York 73
New Zealand Broadcasting Board (NZBB) (1932–36) 121, 132; Plebiscite of Listeners, 1932 136–37, **138**, 139, **140**, 141
New Zealand Broadcasting Service (1936–62) 121, 132
New Zealand Divisional Signal Company 109, 113–14, **114**
New Zealand Graphic and Ladies' Journal 31
New Zealand Listener 126
New Zealand Motion Picture Supplies 161
New Zealand Observer 28, **29**, 31
New Zealand Rugby Union 147
New Zealand, the world's place in 9–10, 22, 83, 84
New Zealand Wireless Troop 109–13, **113**
New Zealand Woman's Weekly 125, 134
Ngata, Apirana 76, 82–83
Ngatea 74, 127, 134, 200
North American Review 28
novelty records 72
N.Z. Radio Record 126, 129–30, 131, 132, 134, 136, 148, 151
N.Z. Record Herald and Kinetoscope News 22, 44, **45**

Ōamaru 38, 104
Oberammergau Passion Play (film) 158–59
O'Callahan, Alice 179
opera 14, 35, 36, 37, **54**, **61**, 62, 132, 169
Opera at Home **54**
Otago Daily Times 93
Otago University 63–64, 116
Otago Witness 39

Paeroa 47, 147; Gaiety Theatre 170
'Painting the Clouds with Sunshine' 191, 193
Pandora 8
Papakura, Maggie 83
Paramount Orchestra **180**
Parlophone Company 76, **77**, 78, 83
Passion Play (film) 158
Pearce, Edna (Aunt Edna, 3YA) 142
Pearse, Arthur 65, 72
'Perfected Phonograph' 26, **29**
performing artists heard on new technologies 14, 36, 37, 38, 46, 47–48, 56, 65; *see also* names of individual performing artists
personal spaces 203
Philco 'Meteor' radio-gramophone **17**
phonautograph 24
phonographs 8, 12, **23**, 30, 52, 201; Auxetophone 38; in cinemas 13, 169–70; connecting people 47; described as musical instruments 38; design and appearance 41, **42**, 43; entertainment 33;

283

exhibitions and demonstrations 21, 22, 24–28, **25**, **29**, 31, 32, 33, 43, 165, 166; international publicity 22; jokes, stories and anecdotes 48–50, **49**; opposition to 38–39; presented as animate and human 39; prices 43; in private households 40–41, **42**, 43, 44–45, **45**, 47, **47**; publicity and guidelines 44, **45**; recitals of recorded music 37–38; revolution in listening 21–22, 24, 27–28, 32–33, 43, 44–48; storage of voices for later remembrance 31–32; unexpected uses 22, 28, 31–32; use in silent films 13, 165–68, **168**; wartime use 50–52, **51**, **52**; *see also* gramophones

pianos 7, 22, 32, 35, 40, 44, 45, 68; impact of phonographs and gramophones 12, 21, 33, 39, 40–41, 43; player pianos 14; prices 43; 'Shut That Old Piano Up' 40–41, 45; silent film music 157, 159, 165, 170, 173, 175, 179–80, 182, 186; superseded by instruments such as guitar 66

Picturedrome cinema, Milford 170, 179, 190

Pierce, George, *Principles of Wireless Telegraphy* 99

HMS *Pioneer* 95

Pipiroa, Hauraki Plains 72

Pitman, Isaac 24

'Pokarekare Ana' 56

popular culture: American, global spread and dominance 14, 18, 132, 193, 194, 195–98; domination of transnational entertainment conglomerates 18; perceived dangers 17, 57; radio 116–17; *see also* global consumer and leisure cultures

Popular Mechanics 99

popular music 204; American, global spread and dominance 14, 18, 132; emotional effects 72; radio 130, 131, 134–35, 136; recorded by the Tahiwis 78; in The Talkeries 36, **37**; between the wars 68; *see also* types of popular music, e.g. dance music

Porourangi, Hohepa 56

portability of sound 8, 9, 12–14, 18, 24, 203, 204; portable gramophones 8, 11, **13**, 53, 72

Post and Telegraph Act 1908 90

Post and Telegraph Amendment Act 1913 90

Post and Telegraph Department 90, 91, 95, 96, 104, 108, 109, 111, 116, 133

preservation of music 14–15

Press (Christchurch) 151; 'Wireless News' 116

Price, Fred and Shirley 143

Priestley, J.B. 204

private homes *see* homes

Progress 97, **98**, 99

public service broadcasting 127, 131–32

public spaces 12, 14, 53, 67, 203

Puechgud, M. 127

Pull Thro', *The* 106

Quo Vadis (film) 177, **178**

radio: durability of sound 16; in the home 123–26, **124**, **125**, 127, 128–29, 135, 139, 141; influence on New Zealand speech 59; learning opportunities for musicians 65; linkages and interaction with television 200–01; mass audience 118; in newspaper columns 116–17, 151; opinions about 89, 124–25, 134–35, 151; perceived noise pollution 134–35; portability of sound 14, 18; repetition of sound 16; separation of sound from sources 8, 9, 12

radio announcers and presenters 137; sports commentators 144, 145; 'uncles' and 'aunts' on children's radio 142, 143–44

Radio Broadcasting Board (1932–36) *see* New Zealand Broadcasting Board (NZBB) (1932–36)

Radio Broadcasting Company (RBC) 120–21, 126, 127, 129, 130, 131–32, 145, 147

radio broadcasting, New Zealand 91, 95, 105, 115–16, 118; British events 150–51; children's programmes 142–44; equipment 116; ex-servicemen 114–15; government administration and control 118, 120–21, 126, 127, 132; Māori performers 148, 150; music 127, 129–32, 134–35, 136, 137, 139, 141; Plebiscite of Listeners, 1932 136–37, **138**, 139, **140**, 141; programming 126–32, 134–35, 136, 137, 139, 141, 142–47; recordings versus live performances 127, 130, 136, 137; remuneration for performers 130–31; serials 122, 127–28, **128**, 147–48, **149**; sports 118, 122, 144–45, **146**, 147; *see also* wireless telegraphy

radio licences 120, 126, 129, 132, 136

radio listening 12–13, 53, 118, 121–23, 135, 139, 204; appointment listening 127–29; communal experience 117–18, 121–22, 150, 151–52; imagination a key element 122, 123, 144, 148, 150, 151,

152; social and cultural contexts 151–52; solitary 148, 151; survey of listeners, 1932 136–37, **138**, 139, **140**, 141; too-loud radio 134–35; unwanted noise from interference 132–33, **133**
Radio New Zealand 132
Radio Record 126, 129–30, 131, 132, 134, 136, 148, 151
Radio Regulations 1925 130
radio sets (receivers) 87–88, 116, 117, **117**, 120, 121, 132; design and materials 121, 123–24, **124**, **125**; technological developments 121; valves 114–15, 121, 137
radio stations: capture of German radio station, Samoa 103, 105–07, **107**, 111; music 8; Sydney 115; Turkish stations, Mesopotamia 112
radio stations, New Zealand 91–92, 116, 117–18, 120–21; 1YA 127, 143, 144, 150; 1ZB 128; 2YA 72, 96, 119–20, 127–28, 132, 137, 142, 143–44, 150; 3YA 127, 142–43, 150; 4YA 127, 143; aerials 91, **92**, 94; B stations, independently owned 121, 126; early government stations 94, 96, 97, **97**, 104; established by Stark, Hicks and Brandon in Dunedin 93–94, **94**; record making 75; silent days 126, 127, 137, 139; YA stations 121, 126
radio transmission 88–89, 116
recitals of recorded music: gramophones **64**, 64–65, 127; phonographs 37–38
recording companies: global industry 34–35, 69, 83, 84; seriousness and respectability 35–38, 43
recordings: availability 33, 43; durability 14–16, 46; familiarity with wide range of music 16; fidelity 46; film songs 191–92, **192**, 194, 199; learning opportunities for musicians 14, 65–66; making recordings 24, **30**, 31–32, 46, 50; New Zealand commercial recordings 75–84; ownership and gifting 15–16; portability 14; repetition of sound 14, 16, 17, 66, 71, 204; separation of sound from sources 8, 9, 12, 52; standard of performers 131, 137, 139; use for learning dances and practising new steps 68, 71–72; *see also* acetate discs; aluminium discs; gramophones; phonographs; resin records; shellac discs; tinfoil; vinyl records; wax cylinders
Reeves, John Sims **29**, 31
Regal recording label 191–92

Reith, John 129, 131
repeatability of sound 12, 16–18, 24, 39, 56, 66, 71, 204
resin records 15
retail trade, New Zealand *see* music retail trade, New Zealand
Rewi's Last Stand/The Stand (film) 189
Rheinhardt, Max 177
Richardson's Entertainers 181
Ridding, Patricia 134
Roe, Eric 185
Rotorua 76, 78, 80
Rotorua Maori Choir 78, 83
Royal Pictures Syndicate 172
Rudd, Steele 147
rugby, radio coverage of matches 144–45, **146**, 147
Russell, Lorraine 124, 150
Russo–Japanese War films 167, 171

Salmon, Mr and Mrs H.A. 33
Salvation Army (New Zealand), films to emphasise messages 162
Samoa, German wireless station 103, 105–07, **107**, 111
San Francisco earthquake, 1906 167, 168
Sandow, Eugen 166
Sartre, Jean-Paul 164
Scannell, Paddy 129
Schafer, Murray 202
Schertzinger, Victor 68
'schizophonia' 202
Scholes, Percy 54, 60
school music 54–58, **55**, **57**
Scott, L.M. 24
Scotti, Antonio 38
serials, radio 122, 127–28, **128**, 147–48, **149**
Shalfoon, Epi 74, 184–85
Shalfoon and the Melody Boys 184, **184**
Shaw, Artie 185
sheet music and scores 14, 66, 136, 181, 194, 199
shellac discs 8, 9, 46, 165
Shelley, James 62
Shepherd, Gwen (Aunt Gwen, 2YA) 119–20, **120**, 142, 143–44
Sheppard, Dawn 127–28, 129
shipping, use of radio 93, 94–95, 96
Shirley, Harry 176, 182
Shirley Intermediate School 58
Sibelius, Jean 60
Sienkiewicz, Henryk 177
silent film music 14, 159, 164–65, 204; arranged scores 177, 179; audience participation 180–82; Brescians 174, 175;

285

choirs 157, 177; classical music 176–77; emotional connection between viewers and images 173; illustrated songs 180–82, **182**; live musicians 165; matched with images 168–69, 175, 179–80, 183; musicians made redundant by talkies 68, 185, 199; orchestras 157, 175–76, 177, 179, **180**; piano 157, 159, 165, 170, 173, 175, 179–80, 182, 186; played separately from films 166, 168; recorded music 165–70, **168**; storage capacities of cylinders and discs 169

silent films: audience noise and behaviour 181–83; combined with variety and vaudeville acts 174–75; fictional narrative films 160; first moving pictures in New Zealand 166; newspaper accounts of technology 158; as novelty 198; popular education and self-improvement 162, 164; roles of the human voice 158–62, **163**, 164, **164**, 173–74; sound effects techniques 171–73, 180, 204; variety of sounds 14, 156, 157, 159, 160, 170, 171

Sinclair, Keith 10

singing 59

Smith, Cob 114

Smith, Walter 191

sound-on-disc and sound-on-film technologies 189

sounds: durability 12, 14–16, 18, 24; fusing with images 155–56; 'hearing the world' 9–10; important role of audio in history, culture and media 156–57; New Zealand engagement with sounds of the world 9–10, 201–02, 203, 204; portability 8, 9, 12–14, 18, 24, 53, 203, 204; repeatability 12, 16–18, 24, 39, 56, 66, 71, 204; separation from sources 8–9, 12, 52, 202; *see also* films; films in New Zealand; hearing; listening; recordings; and entries beginning radio …

Sousa, John Philip, 'The Menace of Mechanical Music' 38–39, 74

South African war films 158, 159, 166–67, 182–83

speakers, radio 134

speech, usage guided by gramophone 58–59

Speedy, Squire 170, 179

Spencer, Len 36

spiritualism and psychic experiences 101–02

Spooner, George 192–93

Sporting Review 28

Spotify 8, 11

St James Theatre, Auckland 191

Stark, Royson 93–94

Stennett, Bruce 119–20

Sterne, Jonathan 202, 204

Stewart, Cal 36

Storm, John 195

Stout, Robert 27

Stowell, Henry H. (Hare Hongi) 148

Strand Theatre, Auckland 68

Stratford, King's Theatre 179

Sullivan, Jim, 'Sounds Historical' programme 143

Sydney 76, 96

Tahiwi, Weno, Hinehou and Henare 78, **79**, 82, 83

Talkeries, The 35–36, **36**, **37**, **47**, 50; marketing 37

'talking machines' 21–22, 31

Tauranga 167, 170

Tayler, E. Douglas 55, 73, 127; *A Complete Scheme of School Music Related to Human Life* 55–57, 58, 59, 60

Taylor & Jones 166

Taylor, Joseph 102

Te Kooti Trail, The (film) 189

Technicolor 191, 192

technology: uses that suit personal lives 22; *see also* audio technologies; digital audio technologies

telegraph 31

telephone lines 100

telephones 22, 31; *see also* mobile phones

television 200; linkages and interaction with radio 200–01

'The Pirate Shop,' Auckland 68

theatres, film *see* cinemas

Thompson, Reginald 112

Thomson, George 197

Three Live Ghosts (film) 191

'Tile Trot' dance 68, **69**, 71, **71**

Timaru 38

Tinakori Hill Morse Wireless Station 91, 96, **97**, 104, 115

tinfoil 8, 22, 24, 26

'Tip-Toe Through the Tulips' 11–12, 15–16, 191, 193

tone tests 65

Toogood, Selwyn 200

Trafalgar Day **164**

Treaty of Waitangi, anniversary of signing 148, 150

Tunohopu Meeting House, Rotorua 78, **80**

Uncle Jack, 3YA (Ernest Bell) 142–43

United States 40; American accent and

INDEX

slang 59, 73, 195–96, 197–98; films and film stars 186, 187, 191, 193–98; global spread and dominance of popular culture 14, 18, 132, 193, 194, 195–98; industrial production of records and films 16, 33, 186, 187, 189; phonograph exhibitions 22, 24; recorded sounds heard in New Zealand 10
Utting, Mary 179

Valentine's Picture Company 170
Vallee, Rudy 74
valves, radio 114–15, 121, 137
Victor Company 35, 41
Victoria College, Wellington 194
Victoria University College 59–60
'Victrola' phonograph 41
vinyl records 7, 8, 165
Vollmöller, Karl 177

Walker, Louis E. 97–99
Wanganui Maori Concert Party 148
Ward, Sir Joseph 90, 93, 101, 108
Waretini, Deane (snr) 76, **77**, 78
Waretini, Deane (jnr), 'Te Piriti' (The Bridge) 76
'warmth' of sound 7, 8
Warner Brothers 191
wartime use of phonographs and gramophones 50–52, **51**, **52**
wax cylinders 8, 9, 15, 25, 26, 32, **36**, 46, 165
'We Parted on the Shore' 181, **182**
Webster, Donald 143
Wellington 108, 134, 175; Bannatyne and Hunter department store 192; Charles Begg & Sons 38, 63; film screenings 192, 198; Jack McEwen's shop, gathering of dance musicians 65; phonograph exhibitions and demonstrations 21, 24–25, **25**, 33; The Talkeries 35, 37; Tinakori Hill Morse Wireless Station 91, 96, **97**, 104, 115
Wellington Centennial Exhibition, 1940 128
Wellington Gramophone Group 63

Wellington Town Hall 38
Welsh, Jack 186–87
West, Thomas James 174
Western Electric 189, 190, 191
West's Pictures 174–75
Whanganui, Opera House 179, 180
Whitehouse, Alfred H. 155, 165–66
Willenborg, Walter J. 93
Williamson's Bio-Tableau film company 171–72
Willochra Tatler 110
Wilson, H.L. 41
wireless telegraphy 87–89; amateur use 87, 88, 89, 90–91, 92–95, 99, 104–05, 115; association with spiritualism and psychic experiences 101–02; cartoons and jokes **100**, 100–01, **101**; earphones 95–96, **97**; first public display 99; first two-way radio conversation between New Zealand and Britain 92; guides and diagrams of technical aspects **98**, 99; New Zealand government use and control 89–91, 94–95, 96, 97; news and magazine reports 96–97, 100, 103; varied early opinions 89, 96–97, 100, 102–03; *see also* military radio; radio broadcasting; radio sets (receivers); radio stations; radio transmission
Wireless Telegraphy Act 1903 87, 89–90
Wodehouse, P.G. 48
'Woman On My Weary Mind' (Bob Crosby) 72
Workers' Educational Association (WEA) 60, 61, 75; Box Scheme 62
World War I 50–52, **51**, **52**, 83, 88, 90–91, 95, 103; Gallipoli 161–62, 163; Mesopotamian Campaign 109–13, **113**; Somme films 182–83; 'The Kid from Timaru' 161–62, **163**; *see also* military radio
writing, use of gramophone accompaniment 58

'Yale Blues' 68

287